D1808001

Ernst Schering Research Foundation Workshop 39
Neuroinflammation – From Bench to Bedside

Springer
Berlin
Heidelberg
New York
Barcelona
Hong Kong
London
Milan
Paris
Tokyo

Ernst Schering Research Foundation
Workshop 39

Neuroinflammation – From Bench to Bedside

H. Kettenmann, G.A. Burton, U.J. Moenning
Editors

With 18 Figures

Springer

Series Editors: G. Stock and M. Lessl

ISSN 0947-6075
ISBN 3-540-43090-3 Springer-Verlag Berlin Heidelberg New York

Die Deutsche Bibliothek - CIP-Einheitsaufnahme
Neuroinflammation - from bench to bedside / H. Kettenmann ... ed.. – Berlin ; Heidelberg ; New York ; Barcelona ; Hongkong ; London ; Milan ; Paris ; Tokyo : Springer, 2002
(Ernst Schering Research Foundation Workshop ; 39)
ISBN 3-540-43090-3

Springer-Verlag Berlin Heidelberg New York
a member of BertelsmannSpringer Science+Business Media GmbH

http://www.springer.de

Typesetting: Data conversion by Springer-Verlag
Printing: Druckhaus Beltz, Hemsbach.
Binding: J. Schäffer GmbH & Co. KG, Grünstadt
SPIN: 10857556 21/3130/AG–5 4 3 2 1 0 – Printed on acid-free paper

Preface

Inflammatory processes have been implicated in the pathophysiology of a variety of neurodegenerative diseases such as Alzheimer's disease, stroke and multiple sclerosis. It is now widely accepted that the brain, considered to be an "immunologically privileged" organ, can exhibit immune and autoimmune responses, and it is these responses that may in fact be responsible for many neural pathologies.

The participants of the workshop

Research over the past decade has demonstrated the pivotal role of the microglia in the neuroinflammatory process, and more recent work in the field of chemokines has firmly established their important role in the pathogenesis of inflammatory disease states.

Inevitably with the wealth of research being conducted in the field of neuroinflammation comes the hope of identifying novel targets for therapeutic intervention in the inflammatory process which may impact on neurodegenerative diseases.

The meeting "Neuroinflammation – From Bench to Bedside" aimed to cover the whole spectrum of research currently being pursued in this exciting field, literally taking the participants from the bench – studying the importance of the microglia in the inflammatory process and the involvement, in particular, of the rapidly expanding chemokine family in the pathophysiology of neurodegenerative diseases – to the bedside – and the possibilities for novel treatment of disorders as diverse as multiple sclerosis, Alzheimer's disease, stroke and variant Creutzfeldt-Jakob disease (vCJD).

H. Kettenmann
G.A. Burton
U.J. Moenning

Table of Contents

1 Cellular Components of Neuroinflammation – An Introduction
H. Kettenmann 1

2 Microglia and the Response to Brain Injury
W.J. Streit 11

3 Elucidating Molecular Mechanisms of Alzheimer's Disease in Microglial Cultures
J. Rogers, L.-F. Lue, D.G. Walker, S.D. Yan, D. Stern, R. Strohmeyer, C.J. Kovelowski 25

4 Neuronal SLC (CCL21) Expression: Implications for Neuron–Microglial Signaling System
K. Biber, A. Rappert, H. Kettenmann, N. Brouwer, S.C.V.M.Copray, H.W.G.M. Boddeke 45

5 Cytokine-Mediated Inflammation and Other Actions in the Central Nervous System
I.L. Campbell 61

6 BSE, Scrapie, and vCJD: Infectious Neurodegenerative Diseases
C. Riemer, D. Simon, S. Neidhold, J. Schultz, A. Schwarz, M. Baier 85

7 Activated Microglia in Alzheimer's Disease and Stroke
J.M. Pocock, A.C. Liddle, C. Hooper, D.L. Taylor, C.M. Davenport, S.C. Morgan 105

8 Inflammation in Stroke – A Potential Target for Neuroprotection?
J. Priller, U. Dirnagl 133

9 Neuroinflammation in Alzheimer's Disease: Potential Targets for Disease-Modifying Drugs
M. Hüll, H. Hampel 159

10 The Concept of In Vivo Imaging of Neuroinflammation with $[^{11}C](R)$-PK11195 PET
A. Cagnin, A. Gerhard, R.B. Banati 180

11 Chemokine Receptors on Mononuclear Phagocytes in the Central Nervous System of Patients with Multiple Sclerosis
C. Trebst, T.L. Sørensen, P. Kivisäkk, M.K. Cathcart, J. Hesselgesser, R. Horuk, F. Sellebjerg, H. Lassmann, R.M. Ransohoff 193

12 The Role of Apoptosis in Neuroinflammation
F. Zipp, O. Aktas, J.D. Lünemann 213

Subject Index 231

Previous Volumes Published in This Series 235

List of Editors and Contributors

Editors

G.A. Burton
Preclinical Drug Research, Schering AG, Müllerstr. 178, 13342 Berlin, Germany (e-mail: Gerardine.Burton@Schering.de)

H. Kettenmann
Max Delbrück Center for Molecular Medicine, Department of Neurosciences, Robert Roessle Str. 10, 13125 Berlin, Germany
(e-mail: Hketten@mdc-berlin.de)

U.J. Moenning
Preclinical Drug Research, Schering AG, Müllerstr. 178, 13342 Berlin, Germany (e-mail: Ursula.Moenning@Schering.de)

Contributors

O. Aktas
Department of Neurology, Division of Neuroimmunology, Charité, Neuroscience Research Center, 10098 Berlin, Germany
(e-mail: orhan.aktas@charite.de)

M. Baier
Robert Koch Institute, Nordufer 20, 13353 Berlin, Germany
(e-mail: BaierM@rki.de)

R. Banati
Hammersmith Hospital, Cyclotron Unit, Du Cane Road, London W12 0NN, UK (e-mail: richard.banti@csc.mrc.ac.uk)

K. Biber
Department of Medical Physiology, University of Groningen, 9700 AD Groningen, The Netherlands (e-mail: k.biber@med.rug.nl)

H.W.G.M. Boddeke
Department of Medical Physiology, University Groningen, A. Densinglaan1, P.O. Box 196, 9712 KZ Groningen, The Netherlands (e-mail: h.w.g.m.boddeke@med.rug.nl

N. Brouwer
Department of Medical Physiology, University of Groningen, A. Deusinglaan 1, P.O. Box 196, 9700 AD Groningen, The Netherlands (e-mail: n.brouwer@med.rug.nl)

A. Cagnin
MRC Cyclotron Unit, Imperial College School of Medicine, Hammersmith Hospital, Du Cane Road, London W12 ONN, UK (e-mail: acagnin@brainaging.it)

I.L. Campbell
Department of Neuropharmacology, The Scripps Research Institute, 10550 North Torrey Pines Road, La Jolla, CA 92037, USA (e-mail: icamp@scripps.edu)

M.K. Cathcart
Department of Cell Biology, The Lerner Research Institute, The Cleveland Clinic Foundation, Cleveland, 9500 Euclid Avenue, OH 44195, USA (e-mail: cathcarm@ccf.org)

S.C.V.M. Copray
Department of Medical Physiology, University of Groningen, A. Deusinglaan 1, P.O. Box 196, 9700 AD Groningen, The Netherlands (e-mail: j.c.v.m.copray@med.rug.nl)

C.M. Davenport
Cell Signalling Laboratory, Institute of Neurology, University College, London, WC1N 1PJ, UK (e-mail: j.pocock@ion.ucl.ac.uk)

U. Dirnagl
Abteilung für Experimentelle Neurologie, Charité, Humboldt-University, Schumannstrasse 20/21, 10098 Berlin, Germany
(e-mail: ulrich.dirnagl@charite.de)

A. Gerhard
MRC Cyclotron Unit, Imperial College School of Medicine, Hammersmith Hospital, Du Cane Road, London W12 ONN, UK
(e-mail: alexander.gerhard@csc.ac.uk)

H. Hampel
Klinikum der Universität München, Klinikund Poliklinik für Psychiatrie und Psychotherapie, Nußbaumstraße 7, 80336 München, Germany
(e-mail: Hampel@psy.med.uni-muenchen.de)

J. Hesselgesser
Department of Immunology, Berlex Biosciences, Richmond, CA, USA
(e-mail: joe_hesselgesser@berlex.com)

C. Hooper
Cell Signalling Laboratory, Institute of Neurology, University College, London, WC1N 1PJ, UK (e-mail: j.pocock@ion.ucl.ac.uk)

R. Horuk
Department of Immunology, Berlex Biosciences, Richmond, CA, USA
(e-mail: richard_horuk@berlex.com)

M. Hüll
Department of Psychiatry, University of Freiburg, Medical School, Hauptstr. 5, 79104 Freiburg, Germany
(e-mail: michael_huell@psyallg.ukl.uni-freiburg.de)

P. Kivisäkk
Department of Neurosciences, The Lerner Research Institute, The Cleveland Clinic Foundation, 9500 Euclid Avenue, Cleveland, OH 44195, USA (e-mail: Kivisap@ccf.org)

C.J. Kovelowski
Sun Health Research Insitute, 10515 W. Santa Fe Drive, Sun City, AZ 85351 USA (email:cjkovelowski@sunhealth.edu)

H. Lassmann
Brain Research Institute, University of Vienna, Vienna, Austria
(e-mail: hans.lassmann@univie.ac.at)

A.C. Liddle
Cell Signalling Laboratory, Institute of Neurology, University College, London, WC1N 1PJ, UK (e-mail: j.pocock@ion.ucl.ac.uk)

L.-F. Lue
Sun Health Research Institute, 10515 W. Santa Fe Drive, Sun City, AZ 85351 USA (e-mail: llue@sunhealth.org)

J.D. Lünemann
Department of Neurology, Division of Neuroimmunology, Charité, Neuroscience Research Center, 10098 Berlin, Germany
(e-mail: jan.lueneman@charite.de)

S.C. Morgan
Cell Signalling Laboratory, Institute of Neurology, University College, London, WC1N 1PJ, UK (e-mail: j.pocock@ion.ucl.ac.uk)

S. Neidhold
Robert Koch Institute, Nordufer 20, 13353 Berlin, Germany
(e-mail: neidholds@rki.de)

J.M. Pocock
Cell signalling Laboratory, Institute of Neurology, University College, 1 Wakefield Street, London WC1NPJ, UK (e-mail: j.pocock@ion.ucl.ac.uk)

J. Priller
Department of Neurology, Charité, Humboldt-University, Schumannstraße 20/21, 10117 Berlin, Germany
(e-mail: Josef.Priller@charite.de)

R. Ransohoff
Department of Neurosciences, NC30, Learner Research Institute, The Cleveland Clinic Foundation, 9500 Euclid Avenue, Cleveland, OH 44195, USA (e-mail: ransohr@ccf.org)

A. Rappert
Max Delbrück Center for Molecular Medicine, Cellular Neuroscience, Robert Roessle Str. 10, 13125 Berlin, Germany (e-mail: arappert@med.rug.nl)

C. Riemer
Robert Koch Institute, Nordufer 20, 13353 Berlin, Germany (e-mail: Riemerc@addcom.de)

J. Rogers
President and Senior Scientist, Sun Health Research Institute, P.O. Box 1278, Sun City, AZ 85351, USA (e-mail: Joseph.rogers@sunhealth.org)

J. Schultz
Robert Koch Institute, Nordufer 20, 13353 Berlin, Germany (e-mail: schultzj@rki.de)

A. Schwarz
Robert Koch Institute, Nordufer 20, 13353 Berlin, Germany (e-mail: schwarza@rki.de)

F. Sellebjerg
Department of Neurology, University of Copenhagen, Glostrup Hospital, Denmark (e-mail: sellebjerg@dadlnet.dk)

D. Simon
Robert Koch Institute, Nordufer 20, 13353 Berlin, Germany (e-mail: simond@rki.de)

T.L. Sørensen
Department of Neurology, University of Copenhagen, Glostrup Hospital, Glostrup, Denmark (e-mail: torbenls@dadlnet.dk)

D. Stern
Columbia University, College of Physicians and Surgeons, 630 West 168th Street, New York, NY 10032, USA (e-mail: dms9@columbia.edu)

W.J. Streit
Department of Neuroscience, P.O. Box 100244, Building 59, University of Florida, College of Medicine, 100 Newell Drive, Gainesville FL 32611, USA (e-mail: streit@ufbi.ufl.edu)

R. Strohmeyer
Sun Health Research Institue, 10515 W. Santa Fe Drive,
Sun City, AZ 85351 USA (e-mail: rstrohmeyer@sunhealth.org)

D.L. Taylor
Cell Signalling Laboratory, Institute of Neurology, University College,
London, WC1N 1PJ, UK (e-mail: j.pocock@ion.ucl.ac.uk)

C. Trebst
Department of Neurosciences, The Lerner Research Institute,
The Cleveland Clinic Foundation, Cleveland OH, USA
(e-mail: trebstc@ccf.org)

D.G. Walker
Sun Health Research Institute, 10515 W. Santa Fe Drive,
Sun City, AZ 85351 USA (e-mail: dgwalker@altavista.com)

S.D. Yan
Columbia University, College of Physicians and Surgeons,
630 West 168th Street, New York, NY 10032, USA
(e-mail: sdy1@columbia.edu)

F. Zipp
Department of Neurology, Division of Neuroimmunology, Charité,
Neuroscience Research Center, 10098 Berlin, Germany
(e-mail: frauke.zipp@charite.de)

1 Cellular Components of Neuroinflammation – An Introduction

H. Kettenmann

1.1 Glial Responses to Brain Injury Were Recognized by Early Pathologists 1
1.2 Microglial Cells Are the Key Elements for the Intrinsic CNS Response to Injury 2
1.3 Astrocytes Assist in the Immune Response 4
1.4 The Blood–Brain Barrier Controls the Access of Immune Cells to the Brain 6
1.5 Inflammatory Responses Influence Brain Diseases 6
References and Suggested Readings 8

1.1 Glial Responses to Brain Injury Were Recognized by Early Pathologists

Inflammation of the central nervous system is a topic which has attracted researchers for more than 150 years. The pathologist Virchow recognized that the brain tissue is not only composed of neurons but contains a second population of cells which he termed glia. His first picture of glial cells, published in 1858 (Virchow 1958), shows cells with morphological features of activated ameboid microglia rather than astrocytes or oligodendrocytes. Since his main focus was the study of pathologic tissue, he may have recognized those glial cells which had responded to inflammation. Subsequently, pathologists described that glial cells surround damaged or dying neurons. In his extensive treatise on "the knowledge of pathologic neuroglia and their relation to degeneration of the CNS" Alois Alzheimer compared glial cells under normal conditions with those of diseased brain. He studied cases of meningitis, dementia, and epilepsy and clearly recognized that damaged and dying neurons were surrounded by ameboid glial cells. He also described that glial cells take over the space which is left after the ganglion neurons had disappeared (Alzheimer 1910). In his drawings it is evident that he had described activated ameboid microglia and reactive astrocytes. Just now we are beginning to understand that these cell types are the two major players for inflammation intrinsic to the CNS and that they exhibit a large repertoire of cellular interactions with the invading immune cells.

1.2 Microglial Cells Are the Key Elements for the Intrinsic CNS Response to Injury

Microglial cells are the intrinsic macrophage population of the central nervous system. They are considered to derive from cells of the monocytic lineage and invade the brain early in development. As early as the 1930s, del Rio-Hortega (1932) recognized focal areas of invasion termed "fountains of microglia". In the central nervous system these cells transform from an ameboid phenotype into a highly ramified form called resting microglia. Microglial cells are abundantly present in all regions of CNS tissue including optic nerve and retina. They seem to occupy a territory in that they do not exhibit physical contact with each other. With their highly ramified morphology, they seem to control a defined volume of brain tissue. While we know little on their function for normal brain activity, their role in pathology related events is well documented, and they can be considered as the major intrinsic CNS component mediating inflammation. Current research focuses on the factors which control the immunological functions of microglial cells and the mechanisms by which microglial cells respond to pathologic insults.

Microglial cells respond with activation to intrinsic pathologic signals such as neurodegeneration, but also to invading pathogens. In many pathologic events and their corresponding animal models the blood brain–barrier is impaired and macrophages from the blood invade the brain parenchyma. Since activated microglial cells acquire many features of macrophages, the two cell populations are difficult to distinguish. It was therefore an important step to develop a model of microglial activation without disturbance of the blood–brain barrier and without macrophage invasion. This paradigm introduced by Kreutzberg's group (Kreutzberg 1996) has become a classical tool: by cutting the facial nerve, motoneurons in the facial nucleus respond to the damage of their axon and in the close vicinity, microglial cells are activated. The advantage of this model is that the blood–brain barrier in the nucleus is not impaired and one can study intrinsic factors which control microglial activation. In the present volume, Georg Kreutzberg and his former collaborator Wolfgang Streit will introduce this model and present results obtained from this experimental paradigm.

Microglial cells can be isolated from the brain and can be studied in cell culture. Much of our knowledge on their cell biological properties results from these in vitro studies. One has, however, to take into account that these cultured cells respond to the unphysiological environment of the culture system and are partially activated which can be recognized by their ameboid morphology. This has to be taken into account when extrapolating data obtained in culture to the situation in vivo. An important question addressed in this volume involves the signals by which microglial cells sense the pathologic environment. Microglial cells express neurotransmitter receptors including those for ATP, AMPA, and substance P and thus have the capacity to detect neuronal activity. How the cells in the tissue environment respond to these neurotransmitters is not yet well described, since all the studies were obtained from cultured cells. Yet, there are more candidates, which may signal neuronal or neural injury to microglial cells. There is increasing evidence that cytokines are potential signals from damaged neurons to microglial cells. One candidate is the cytokine CCL21, which has been demonstrated by Erik Boddeke's group to be released from injured neurons under ischemic conditions. It has been shown by a collaboration of our and Boddeke's group that microglial cells express receptors for CCL21 and respond with the activation of a chloride current and cell migration to this injury. Interestingly, the chloride channel activity is a prerequisite for cell migration. Another group of candidates for neuron-microglial signaling are neurotrophins. Microglial cells also respond to signals which invade the brain, such as viruses or bacteria. Lipopolysaccharide (LPS), a component of the membrane of gram-negative bacteria, is a prominent stimulator of microglial activation. LPS binds to CD14 and its binding is further regulated by the presence of LPS-binding protein. Cell wall components of gram-positive bacteria also trigger microglial activation. There are a number of other receptors which are understood to control microglial activation, such as complement receptors or the many different cytokine receptors expressed by microglial cells. The disease-associated form of the prion protein is also an activator of microglial cells as elaborated in the chapter by Michael Baier and colleagues.

Upon activation, microglial cells will profoundly affect the local inflammatory response by secreting a variety of different cytokines such as interleukin (IL)-1, tumor necrosis factor (TNF)-α, IL-12, IL-18. This release pattern will strongly influence the surrounding cells, namely the intrinsic CNS populations, neurons, astrocytes, and oligodendrocytes as well as the invading cells from the blood. These complex interactions are more clearly described in the chapters by Ian Campbell and Wolfgang Streit. Microglial cells also express a large variety of different chemokines and their respective receptors and thereby are a part of the chemokine signaling pathway. An important aspect of microglial function is the presentation of antigens. Microglial cells are the most prominent antigen-presenting cells of the central nervous system. While they only purely express MHC class II under normal conditions the expression of this important molecule for antigen presentation is upregulated under inflammatory and neurodegenerative conditions. In concert with a number of growth stimulatory molecules such as for instance B7-1, microglial cells are therefore the partners which interact with the invading T cells. Microglial cells are thus central cellular elements which control the extent of neuroinflammation.

1.3 Astrocytes Assist in the Immune Response

The second important intrinsic CNS player mediating neuroinflammation are the astrocytes. While the function of microglial cells under normal physiological conditions is not known, there are a number of putative functions for astrocytes. This includes guidance of neuronal migration, maintenance of the extracellular environment, and energy supply for neurons. In the normal brain, astrocytes can exhibit a quite diverse morphology such as the radial structure of Bergmann glial cells or a more compact appearance in the cortex. A common feature of all astrocytes is their contact with the neuronal membrane on one side and with non-brain structures such as the pial surface or capillaries on the other. In response to pathologic events, astrocytes undergo morphological and functional changes, and this process is termed astrogliosis. Astrocytes in pathologic tissue upregulate the astrocyte-specific protein, glial fibrillary acidic protein, and the cells are termed reactive astrocytes. In cell culture, astrocytes also prominently express glial fibrillary acidic protein, and thus the cultured astrocytes can be viewed as a model for the reactive astrocytes in the brain. This always has to be taken into account in comparing cultured data to those in vivo.

Astrocytes have the capacity to produce a large variety of cytokines and chemokines both in vitro and in vivo. This includes for example IL-6, interferon β, TNF-α, transforming growth factor (TGF)-β and the monocyte chemoattractant protein 1. They thus release products which will affect microglial responses. The importance of cytokine release during injury and its impact on the process of neuroinflammation is addressed in the chapter by Ian Campbell. While astrocytes in vitro can function as antigen-presenting cells, there are still conflicting data about extent to which astrocytes function as antigen-presenting cells in vivo. This refers to MHC class II expression as well as costimulatory molecules such as B7-1 or B7-2. Yet intercellular adhesion molecule (ICAM)1, an important factor for antigen presentation, is expressed by astrocytes. Recent evidence indicates that astrocytes function as non-professional antigen-presenting cells. This implies that they promote mainly T helper (TH)2 responses thus stimulating a subpopulation of the $CD4^+$ TH cells. TH2 cells produce the cytokines IL-4, IL-5, IL-6, IL-10, and IL-13 and thus are important for downregulating macrophage activation. Moreover, astrocytes do not stimulate T cell proliferation but, in contrast, support apoptosis for $CD4^+$ T cells (for review see Dong and Benveniste 2001). Thus, astrocytes are part of the network of the immune effector cells in the brain by interacting with both microglia and with the invading cells from the periphery.

1.4 The Blood–Brain Barrier Controls the Access of Immune Cells to the Brain

The blood–brain barrier controls the access of peripheral immune cells to the central nervous system. Under normal conditions this barrier prevents the access of peripheral cells and only rarely cells from the blood such as T cells spontaneously invade the brain parenchyma. The barrier is formed by the endothelial cells which are interconnected by tight junctions. Brain endothelial cells differ from endothelial cells in other tissues where they are present as fenestrated, leaky endothelial cells. Astrocytes are involved in the formation of this barrier in that they are thought to deliver the signal to the endothelial cells for the formation of these tight contacts. Endothelial cells are contacted by the astrocyte endfeet which form an ensheathment around the endothelium. There are also contacts of microglial cells to the endothelial membrane. The perivascular space around microvessels in the brain is filled by basement membrane that is secreted by both glial and endothelial cells. The interaction of microglia, astrocytes and endothelial cells will determine the tightness of this barrier and thus controls the entry of leukocytes into the brain and thus the extend of the inflammatory process in the central nervous system. Like astrocytes and microglial cells, endothelial cells can express and release chemokines and cytokines and thereby influence the inflammatory process. They are, moreover, the first cells to interact with the invading leukocytes, e.g. by the expression of the cell adhesion molecules ICAM or vascular cell adhesion molecule (VCAM).

1.5 Inflammatory Responses Influence Brain Diseases

Diseases of or injury to the CNS will lead to the death of neurons and thus will lead to the impairment of brain function. It is well established that neurodegeneration is accompanied by responses of the non-neuronal cells and thus all these neuropathologic events have an inflammatory component. In the following chapters the cellular responses of the immune-relevant cells to major diseases, namely Alzheimer's disease, stroke, multiple sclerosis, and transmissible spongiform encephalopathies, are reviewed. Microglial cells express and respond to proteins relevant in Alzheimer's disease and this topic is reviewed in the chapter by Joseph Rogers and colleagues based on a series of experiments on adult human microglia. This chapter is complemented by the review of Michael Hüll and Harald Hampel which focuses on potential targets of drugs to modify inflammatory responses in Alzheimer's disease. The importance of activated microglial cells for Alzheimer's disease is further highlighted in the review by Jennifer Pocock and the responses are compared with inflammatory events observed in stroke. While stroke was long considered as an acute event, it becomes evident that the acute phase is followed by a chronic phase which has many features of an inflammatory event. The review by Josef Priller and Ulrich Dirnagl explores the potential in modulating this late phase with the goal to reduce neurodegeneration. A neurologic disorder which has been long connected with inflammation is multiple sclerosis. The review by Richard Ransohoff and colleagues summarizes the expression profile of cytokines by mononuclear phagocytes (intrinsic microglial and invaded macrophages) in areas of demyelination and thus define possible targets for drug interference. The response of the immune elements in transmissible spongiform encephalopathies such as bovine spongiform encephalopathy, scrapie, or the new variant of the Creutzfeldt-Jakob disease are summarized in the review of Michael Baier and colleagues. A new approach of tracing activated microglial cells in humans has been explored by Richard Banati. By using a ligand for the peripheral benzodiazepine receptor which is preferentially expressed by activated microglia, it becomes possible to visualize areas of inflammation in human patients using positron emission tomography (PET). All these reviews demonstrate that pathologic events in the brain are accompanied by an immunological component. Yet our understanding of this complicated interplay between all these cellular partners is incomplete and it seems impossible to predict what will happen if one interferes, with drugs, at one point within this cascade.

References and Suggested Readings

Abbott NJ (2000) Inflammatory mediators and modulation of blood-brain barrier permeability. Cell Mol Neurobiol 20:131–147

Aloisi F (2001) Immune function of microglia. Glia 36:165–179

Aloisi F, Ria F, Adorini L (2000b) Regulation of T-cell responses by CNS antigen-presenting cells: different roles for microglia and astrocytes. Immunol Today 21:141–147

Alzheimer A (1910) Beiträge zur Kenntnis der pathologischen Neuroglia und ihre Beziehung zu den Abbauvorgängen im Nervengewebe. In: Nissl F, Alzheimer A (eds) Histologische und Histophatologische Arbeiten über die Grosshirnrinde mit besonderer Berücksichtigung der Pathologischen Anatomie der Geisteskrankheiten. Verlag von Gustav Fischer, Jena, pp 401–562

Bauer J, Rauschka H, Lassmann (2001) Inflammation in the nervous system – the human perspective. Glia 36:235–243

Becher B, Prat A, Antel JP (2000) Brain-immune connection: immuno-regulatory properties of CNS-resident cells. Glia 29:293–304

Benn T, Halfpenny C, Scolding N (2001) Glial cells as targets for cytotoxic immune mediators. Glia 36:200–211

Del Rio-Hortega P (1932) Microglia. In: Penfield W (ed) Cytology and cellular pathology of the nervous system. Paul P. Hocker Inc., New York, pp 481–534

Dong Y, Benveniste EN (2001) Immune function of astrocytes. Glia 36:180–190

Flügel A, Bradl M (2001) New tools to trace populations of inflammatory cells in the CNS. Glia 36:125–136

Glabinski AR, Ransohoff RM (1999) Chemokines and chemokine receptors in CNS pathology. J Neurovirol 5:3–12

Hatten ME, Liem RKH, Shelanski ML, Mason CA (1991) Astroglia in CNS injury. Glia 4:233–243

Hickey WF (1999) Leukocyte traffic in the central nervous system: the participants and their roles. Semin Immunol 11:125–137

Hickey WF (2001) Basic principles of immunological surveillance of the normal central nervous system. Glia 36:118–124

Iglesias A, Bauer J, Litzenburger T, Schubart A, Linington C (2001) T and B cell responses to myelin oligodendrocyte glycoprotein in experimental autoimmune encephalomyelitis and multiple sclerosis. Glia 36:220–234

Kettenmann H, Ransom BR (1995) Neuroglia. Oxford University Press, New York

Kreutzberg GW (1996) Microglia: a sensor for pathological events in the CNS. Trends Neurosci 19:312–318

Miller DW (1999) Immunobiology of the blood-brain barrier. J Neurovirol 5:570–578

Neumann H (2001) Control of immune function by neurons. Glia 36:191–199

Owens T, Wekerle H, Antel J (2001) Genetic models for CNS inflammation. Nat Med 7:161–166

Pender MP, Rist MJ (2001) Apoptosis of inflammatory cells in immune control of the nervous system – the role of glia. Glia 36:137–144

Prat A, Biernacki K, Wosik K, Antel JP (2001) Glial cell influence on the human blood-brain barrier. Glia 36:145–155

Stoll G, Jander S (1999) The role of microglia and macrophages in the pathophysiology of the CNS. Prog Neurobiol 58:233–247

Streit WJ, Walter SA, Pennell NA (1999) Reactive microgliosis. Prog Neurobiol 57:563–581

Virchow R (1858) Die Cellularpathologie in ihrer Begründung auf physiologische und pathologische Gewebelehre. Verlag von August Hirschwald. Berlin

Williams K, Alvarez X, Lackner AA (2001) Central nervous system (CNS) perivascular cells are immunoregulatory cells that connect the CNS with the peripheral immune system. Glia 36:156–164

2 Microglia and the Response to Brain Injury

W.J. Streit

2.1 Introduction and Overview 11
2.2 Functional Plasticity of Microglia 13
2.3 Functional Significance of Microglial Activation 14
2.4 Neuron–Microglia Communication 16
2.5 Microglia as the Brain's Immune System 18
2.6 Microglia in Alzheimer's Disease 19
References 20

2.1 Introduction and Overview

Much has been learned about microglial cells in recent years, and perhaps the most consistent and important observation made in many different laboratories is that microglia are the first non-neuronal cells that respond to CNS injury. Microglia have been appropriately called "sensors of pathology" because of their ability to react quickly to virtually all kinds of acute CNS injuries (Kreutzberg 1996). The response of microglia to CNS injury is called microglial activation, and this process involves no less than four distinct characteristics of the microglial cell: its morphology, its mitotic activity, its surface phenotype, and its secretory activity. Depending on the nature and severity of the injury, these cellular characteristics may be influenced differentially. For example, microglial mitosis is greater after acute necrosis of motoneurons than after axotomy (Streit and Kreutzberg 1988), and microglial secretion of proinflammatory cytokines, interleukin (IL)-1β and tumor necrosis factor (TNF)-α, is greater after traumatic spinal cord injury than after motoneuron axotomy (Streit et al. 1998). Thus, microglial activation represents a multi-faceted cellular phenomenon that generates many, and functionally different, microglial activation states.

What causes microglial activation? In vivo studies clearly support the notion that microglial activation is triggered by neuron-derived signals, because it is limited to those anatomical regions that are occupied by neurons affected by the injury. Importantly, neuronal injury does not need to result in obvious structural damage for it to trigger microglial activation. Even mild damage, or perhaps merely a change in neuronal activity, can generate sufficient signals that may set off some aspect of microglial activation. One example of a mild neuronal disturbance that causes primarily a change in the microglial surface phenotype is cortical spreading depression (Gehrmann et al. 1993). Since there are several components to the microglial activation program, and since these are differentially affected by various types of CNS injuries, it is reasonable to hypothesize that neurons can generate multiple signals, each regulating distinct aspects of the microglial activation program. Elucidating the nature and activity of neuronal signals involved in the regulation of microglial activation is the subject of ongoing research and two specific candidate molecules, IL-6 and fractalkine, are discussed below. A better understanding of the molecular interactions that take place between neurons and microglia will be essential for continued clarification of the pathogenetic mechanisms that underlie CNS injury and disease.

Signals that are produced by injured or otherwise perturbed neurons likely include molecules that recognize and bind to appropriate receptors on the microglial cell surface. Microglia are well equipped to receive and process multiple paracrine signals because they carry a great variety of receptor molecules. These include ion channels, as well as receptors for nucleotides, neuropeptides, chemokines/cytokines, growth factors, and neurotransmitters (Streit 2002). In addition, microglia have receptors for certain serum components, such as complement, thrombin, or immunoglobulins (Graeber et al. 1988a; Ulvestad et al. 1994; Müller et al. 2000). Other recognition molecules, known to be present on the surface of peripheral immune cells, such as CD4 and CD8 co-receptors, major histocompatibility complex (MHC) antigens, integrins, and surface immunoglobulin add to this, already diverse, microglial surface jungle of adhesion molecules and receptors. It remains a challenge to determine how exactly the multitude of extracellular signals are processed by microglia and how these signals are efficiently transduced into gene transcription events that ultimately define the microglial response to neuronal injury. Microglial signal transduction events will undoubtedly continue to be a major focus of future research.

2.2 Functional Plasticity of Microglia

The concept of microglial functional plasticity is based on morphological observations describing the ability of microglia to transform from a ramified to a reactive to a rounded cell shape (Streit et al. 1988). The observations were made in the rat facial nucleus where microglia respond vigorously to lesioning of the facial nerve. Crushing or cutting the facial nerve elicits a retrograde reaction in facial motoneurons which is accompanied by microglial activation that encompasses all of the aforementioned characteristics: morphology, proliferation, surface phenotype, and cytokine/growth factor production. Microglial cells become enlarged by retracting their highly branched cell processes and surround axotomized motoneurons, they divide to increase their numbers, they alter their surface phenotype either by upregulating the expression of constitutively expressed surface molecules or by expressing new ones, and they produce increased amounts of cytokines/growth factors, in particular, transforming growth factor (TGF)-β (Graeber et al. 1988a,b; Streit et al. 1989; Kiefer et al. 1993). This classical microglial response to neuronal injury supports the broader idea that it is injured neurons, in general, that stimulate the various changes associated with microglial activation. Since axotomized motoneurons do regenerate, the rapid triggering of microglial activation likely occurs because injured neurons are recruiting nearby microglia to assist them in their struggle to survive and to regenerate. The ensuing abutment of microglial and neuronal cell bodies provides a spatial arrangement that allows for a number of neuron–glia interactions to occur. Perineuronal microglia displace synaptic terminals (Blinzinger and Kreutzberg 1968), perhaps to minimize afferent stimulation of axotomized neurons. Perineuronal microglia may synthesize and release trophic molecules, such as TGF-β, to help sustain motoneuron survival. Should some motoneurons nonetheless fail to recover from injury and die, perineuronal microglial cells are already perfectly situated to commence with phagocytosing the dead neuron. If there is neuronal death, microglia transform from an activated to a phagocytic state assuming the rounded, or ameboid, shape that is characteristic of brain macrophages. Thus, functional plasticity of microglia refers to a continuum of morphological transitions of microglial cell shape from the highly branched (resting) phenotype to the full-fledged, rounded morphology of a brain macrophage. Different functions of the various morphological intermediates are reflected in the cells' activities as perineuronal neuroprotective effectors, as synapse-remodeling elements, and as phagocytes. The changes in cell shape are accompanied by changes on the cell surface that may involve enhanced expression of some constitutive surface proteins, but can also include de novo expression of surface antigens on activated cells that are absent from non-activated ones. As with the changes in morphology, the development of different patterns of surface phenotypes represent a continuum of changes such that certain patterns may reflect certain functions. The best example of one such correlation between phenotype and function is the expression of MHC class II antigens on microglia, which has suggested that microglia function as the antigen-presenting cells of the CNS (Hickey and Kimura 1988).

As mentioned, the concept of microglial functional plasticity arose from a limited number of studies conducted in the facial nerve paradigm, but subsequent work has shown that the morphological and phenotypic changes on microglia occur with remarkable consistency in other models of acute brain injury and disease, such as ischemia, spinal cord injury, malignant brain tumors, and neurotoxicant-induced brain damage (Morioka et al. 1991, 1992; McCann et al. 1996; Streit et al. 1998). The implication is that with the appearance of certain microglial phenotypes in these brain lesions some of the same functional activities are being expressed.

2.3 Functional Significance of Microglial Activation

There has been much debate about whether microglia are neuroprotective or neurotoxic, and I will not discuss the pros and cons of this issue in great detail. In brief, there is very little evidence from in vivo studies in acute CNS injury to suggest that microglia are harmful cells. This may be different for neurodegenerative diseases, such as Alzheimer's disease, and I will consider this topic separately below. For purposes of the present discussion, it may be useful to briefly summarize some in vivo observations that support a neuroprotective role and argue against a neurotoxic function of microglial cells. First is a point already made above, namely that post-axotomy survival and subsequent axonal regeneration of motoneurons is accompanied by vigorous microglial activation. Therefore, it would be reasonable to argue that microglial activation, which begins long before axons have regenerated, aids in the regeneration of injured motoneurons. The activated microglial cells that surround and ensheath injured perikarya may have pro-regenerative roles analogous to those of peripheral macrophages at the axotomy site (Perry et al. 1987). Second is the observation that, in contrast to a peripheral axotomy, a central axotomy, such as transection of the rubrospinal tract in the cervical spinal cord, elicits only minimal microglial activation that is evident as slightly hypertrophic microglial cells in the red nucleus (Barron et al. 1990; Tseng et al. 1996; Streit et al. 2000). Rubrospinal tractotomy does not elicit microglial mitosis, or perineuronal ensheathment of neurons by reactive microglia, and it results in neuronal atrophy. This supports the idea that axotomized rubrospinal neurons generate insufficient signals to stimulate robust perineuronal glial inflammation that may help prevent their atrophy. Third, when microglial activation occurs in acute brain injuries where selective neuronal vulnerability plays a role, such as global cerebral ischemia, there is no evidence that additional (or secondary) neuronal damage occurs because activated microglia are present. Following four-vessel occlusion, CA1 neurons undergo delayed neuronal death, whereas immediately adjacent CA2 neurons survive, and there is an abrupt transition from dying to surviving neurons and from presence to absence of activated microglia at the CA1/CA2 border (Morioka et al. 1991). If activated microglia were indeed harmful, one would expect to see evidence of microglial neurotoxicity in the CA2 region, which is not the case. Moreover, if delayed death of CA1 neurons is prevented by administration of the anti-excitotoxic agent, MK-801, microglial activation is attenuated showing that preventing neuronal damage results in less microglial activation (Streit et al. 1992). This further supports the notion that intercellular signaling between neurons and microglia is primarily unidirectional, from neuron to microglia.

One last point concerns the production of potentially neurotoxic substances by microglial cells in vivo. Of particular interest in this regard are proinflammatory cytokines, IL-1β and TNF-α, molecules that are known to be produced by reactive microglia and that are considered potentially damaging within the context of a chronic inflammatory reaction. In recent years, it has become possible to measure mRNA and protein of these and other cytokines in tissue samples from the acutely injured CNS. Consistently, such measurements have shown that production of IL-1β and TNF-α is a transient event beginning almost immediately after the injury, rising to a peak within hours, and then declining to basal levels within a day or two (Streit et al. 1998; Pearson et al. 1999). These observations suggest that production of IL-1β and TNF-α is strictly controlled limiting potentially detrimental effects caused by prolonged presence of these cytokines. At the same time, the transient but powerful induction of IL-1β and TNF-α immediately after injury could serve as a trigger for secondary changes, such as inducing synthesis of growth factors that may be beneficial for post-traumatic repair processes. In other words, immediate but brief post-injury inflammation is likely to be required to initiate the wound healing process in the CNS, as it does in peripheral tissues.

2.4 Neuron–Microglia Communication

If neurons control the various aspects of microglial activation, what molecules are involved in mediating the signaling between neurons and microglia? Although of critical importance for a fundamental understanding of CNS disease and injury, this area of research is still in its infancy and few candidate signaling molecules have been identified. Recent research has pointed towards IL-6 and the chemokine fractalkine as potential signaling agents that control distinct facets of the microglial activation program. Based on cell culture studies, IL-6 together with IL-1β and TNF-α have long been considered proinflammatory mediators that are produced by activated glial cells (e.g., Hetier et al. 1988; Gottschall et al. 1995). These in vitro observations stand in contrast to studies conducted in vivo that consistently report a neuronal localization of this cytokine (Lemke et al. 1998; März et al. 1998; Murphy et al. 1995, 1999). In addition, studies from this and other laboratories have shown that there is a strong correlation between the induction of IL-6 mRNA expression and neuron regeneration supporting the idea that IL-6 is a pro-regenerative and neuroprotective factor (Hirota et al. 1996; Zhong et al. 1999; Streit et al. 2000). The concept that has emerged specifically from studies in the facial nerve system is that injured motoneurons produce increased amounts of IL-6 in order to help stimulate the proliferative burst that is part of microglial activation in the axotomized facial nucleus. The evidence for this is as follows: first is the time course of IL-6 mRNA expression after motoneuron injury. This shows a rapid rise hours after injury and well before the onset of microglial proliferation. Levels of IL-6 mRNA then decline gradually during the first week after injury when microglial proliferation occurs. Incidentally, IL-6 mRNA expression does not occur in axotomized rubrospinal neurons. Second is the observation that cultured microglia express IL-6 receptors and respond to exogenously added IL-6 with increased proliferation. Third is the observation that in IL-6-deficient mice there is reduced glial proliferation (Klein et al. 1997). Of course, IL-6 may not act alone in triggering microglial proliferation but could interact with other, better-known stimulators of microglial mitosis, such as granulocyte-macrophage colony stimulating factor (GM-CSF) whose mRNA levels remain unchanged after facial nerve axotomy (Raivich et al. 1998; Streit et al. 1998). In addition, it is conceivable that IL-6 may exert some influence over the development of the surface phenotype of activated microglia, or over microglial cytokine/growth factor production, but such roles have not yet been investigated. As a working hypothesis, it is suggested that IL-6 exerts its neuroprotective role by being involved in the recruitment of reactive glial cells.

In contrast to IL-6, the currently perceived role of fractalkine in regulating reactive microgliosis is quite different (Streit et al. 2001). Like IL-6, fractalkine mRNA is found in neurons and its receptor, CX3CR1, is present on microglia when localized in tissue sections and in vitro (Harrison et al. 1998; Nishiyori et al. 1998; Schwaeble et al. 1998). These observations suggest that neuron-microglia communication may in part be mediated through fractalkine and CX3CR1. Although the precise nature of this interaction remains to be defined, a comprehensive evaluation of current in vivo and in vitro data suggests that fractalkine's role is related to providing some inhibitory control over microglial activation. When motoneurons are axotomized, levels of fractalkine mRNA decrease in the injured neurons at the same time that microglial activation occurs, suggesting that a lifting of constitutive inhibition allows microglial activation to proceed (Harrison et al. 1998). Fractalkine receptor-deficient mice show an apparently normal microglial proliferative response in the axotomized facial nucleus (Jung et al. 2000), showing that fractalkine is not involved in the regulation of microglial proliferation. When cultured microglia are stimulated with lipopolysaccharide (LPS) to produce TNF-α and to become neurotoxic, addition of fractalkine attenuates both TNF production and neurotoxicity (Zujovic et al. 2000). Taken together, these observations suggest that fractalkine has constitutive anti-inflammatory action in the CNS by controlling microglial production of proinflammatory cytokines. Fractalkine-mediated inhibitory action may be one of several mechanisms acting in synergy to maintain an inhibitory microenvironment in the CNS. Recent findings by Hoek et al. (2000) have identified a neuronal membrane glycoprotein, OX2 (CD200) that serves a similar purpose.

2.5 Microglia as the Brain's Immune System

The idea that microglia form a network of immunocompetent cells in the CNS has succeeded the notion of the brain as an immunologically privileged organ without rendering the latter obsolete. It has helped sharpen the concept by stating that the brain's immune privilege is not absolute, but subject to a kind of internal immune-surveillance system consisting of microglia as the key cellular component and is thus independent of blood-borne leukocytes. The major observations that have contributed to the idea that microglia represent the brain's internal immune system include the detection of MHC antigens, T and B lymphocyte markers, and other immune-cell antigens on microglia and the fact that microglia, in general, have properties similar to monocytes and other cells of the macrophages lineage. Among such properties is their ability to produce cytokines typically produced by immune accessory cells, and to serve as antigen-presenting cells (APCs) (Frei et al. 1987; Hetier et al. 1988; Hickey and Kimura 1988). Although it is clear that of all the CNS parenchymal cell types microglia are the most likely to function as APCs, there are caveats. As it turns out, microglia are only weak APCs compared to peripheral APCs from other organs, as shown in a number of studies from different laboratories (Ford et al. 1995; Carson et al. 1998, 1999; Flügel et al. 1999). This is perhaps not surprising since the CNS also harbors other macrophages that, although not part of the parenchyma, populate the critical interface between the CNS parenchyma and the blood. These cells are known as perivascular cells because they reside in the perivascular space (Graeber and Streit 1990), and they may be considered "professional" macrophages, since they are bone marrow-derived and capable of presenting antigen strongly (Hickey and Kimura 1988; Ford et al. 1995). These so-called "other CNS macrophages" (Ford et al. 1995) represent a subset that is phenotypically distinct ($CD11b/c^{+}$ and $CD45^{hi}$) from parenchymal microglia (which are $CD11b/c^{+}$ and $CD45^{low}$), and they have immunological properties of peripheral accessory cells. Because of their location in the perivascular space, these cells are likely to come into contact with and process CNS antigens that may seep from the parenchyma into the perivascular space, and they may present these CNS antigens to T lymphocytes. Microglia, on the other hand, reside in the CNS parenchyma where they do not usually encounter T cells, and hence there may not be a need for microglia to present antigen within the CNS microenvironment. When a traumatic brain lesion occurs and the blood–brain barrier becomes disrupted, white cells from the blood, including APCs and lymphocytes, gain access to the brain parenchyma and to brain antigens that could then be presented to lymphocytes by extravasating peripheral APCs. That T cells do, in fact, become sensitized to brain antigens is suggested by studies which have shown that T cells isolated from spinal-injured rats are capable of causing histopathologic changes similar to experimental autoimmune encephalomyelitis (EAE) when injected into naïve animals (Popovich et al. 1996).

2.6 Microglia in Alzheimer's Disease

Chronic inflammation is thought to play a role in the pathogenesis of Alzheimer's disease (AD) (Akiyama et al. 2000), but it is important to note that inflammation in the AD brain is restricted largely to endogenous CNS cells, primarily microglia, but also astrocytes. Unlike in

EAE or multiple sclerosis (MS), where mononuclear leukocytes invade the CNS parenchyma in great numbers, the presence of lymphocytes and other blood-borne leukocytes in AD brain is the exception rather than the norm. Thus, in AD there is chronic glial neuroinflammation which is characterized by widespread glial fibrillary acidic protein (GFAP) immunoreactivity in astrocytes, as well as by IL-1α and MHC II immunoreactivity in microglia. Morphological signs of microglial activation are also apparent, and microglia are generally enlarged and form numerous clusters at sites where amyloid-β (Aβ) protein has been deposited. It is known that microglia are able to clear Aβ (Frautschy et al. 1992; Schenk et al. 1999), and the clustering of microglia around Aβ deposits in AD brain likely occurs because the cells are trying to remove the insoluble protein deposits. However, clearance of Aβ is not achieved in the AD brain and this raises the possibility that the Aβ-clearing ability of microglia may be weakened or lost. In addition, there may be overproduction of Aβ such that microglia are being overwhelmed by a greater-than-normal amyloid burden. This may compromise not only the cells' Aβ-clearing ability, but could also impact other cell functions, such as neuroprotection and trophic functions to aid in the survival of neurons weakened by old age or otherwise. If dysfunctional microglia are neglecting their normal duties related to sustaining neuronal well-being, this could ultimately result in the development of neurodegenerative changes, such as neurofibrillary tangle formation. Thus, an alternative view to the inflammatory hypothesis of AD pathogenesis could be that neurodegenerative changes occur because microglia are becoming dysfunctional, or diseased, and are therefore neglecting neurons rather than attacking them. If microglia are indeed becoming dysfunctional, there should be structural and/or biochemical evidence for diseased microglia. Studies are underway in this laboratory to investigate whether microglia themselves show signs of degenerative changes.

References

Akiyama H, Barger S, Barnum S, Bradt B, Bauer J, Cole GM, Cooper NR, Eikelenboom P, Emmerling M, Fiebich BL, Finch CE, Frautschy S, Griffin WST, Hampel H, Hull M, Landreth G, Lue LF, Mrak R, Mackenzie M, O'Banion K, Pachter J, Pasinetti G, Plata-Salaman C, Rogers J, Rydel R,

Shen Y, Streit W, Strohmeyer R, Tooyoma I, Van Muiswinkel FL, Veerhuis R, Walker D, Webster S, Wegrzyniak B, Wenk G, Wyss-Coray A (2000) Inflammation and Alzheimer's disease. Neurobiol Aging 21:383–421

Barron KD, Marciano FF, Amundson R, Mankes R (1990) Perineuronal glial responses after axotomy of central and peripheral axons. A comparison. Brain Res 523:219–229

Blinzinger K, Kreutzberg G (1968) Displacement of synaptic terminals from regenerating motoneurons by microglial cells. Z Zellforsch Mikrosk Anat 85:145–157

Carson MJ, Sutcliffe JG, Campbell IL (1999) Microglia stimulate naïve T-cell differentiation without stimulating T-cell proliferation. J Neurosci Res 55:127–134

Carson MJ, Reilly CR, Sutcliffe JG, Lo D (1998) Mature microglia resemble immature antigen-presenting cells. Glia 22:72–85

Flügel A, Labeur MS, Grasbon-Frodl EM, Kreutzberg GW, Graeber MB (1999) Microglia only weakly present glioma antigen to cytotoxic T cells. Int J Dev Neurosci 17:547–556

Ford AL, Goodsall AL, Hickey WF, Sedgwick JD (1995) Normal adult ramified microglia separated from other central nervous system macrophages by flow cytometric sorting. Phenotypic differences defined and direct ex vivo antigen presentation to myelin basic protein-reactive $CD4^+$ T cells compared. J Immunol 154:4309–4321

Frautschy SA, Cole GM, Baird A (1992) Phagocytosis and deposition of vascular β-amyloid in rat brains injected with Alzheimer β-amyloid. Am J Pathol 140:1389–1399

Frei K, Siepl C, Groscurth P, Bodmer S, Schwerdel C, Fontana A (1987) Antigen presentation and tumor cytotoxicity by interferon-gamma-treated microglial cells. Eur J Immunol 17:1271–1278

Gehrmann J, Mies G, Bonnekoh P, Banati R, Iijima T, Kreutzberg GW, Hossmann KA (1993) Microglial reaction in the rat cerebral cortex induced by cortical spreading depression. Brain Pathol 3:11–17

Gottschall PE, Yu X, Bing B (1995) Increased production of gelatinase B (matrix metalloproteinase-9) and interleukin-6 by activated rat microglia in culture. J Neurosci Res 42:335–342

Graeber MB, Streit WJ (1990) Perivascular microglia defined. Trends Neurosci 13:366

Graeber MB, Streit WJ, Kreutzberg GW (1988a) Axotomy of the rat facial nerve leads to increased CR3 complement receptor expression by activated microglial cells. J Neurosci Res 21:18–24

Graeber MB, Tetzlaff W, Streit WJ, Kreutzberg GW (1988b) Microglial cells but not astrocytes undergo mitosis following facial nerve axotomy. Neurosci Lett 85:317–321

Harrison JK, Jiang Y, Chen S, Xia Y, Maciejewski D, McNamara RK, Streit WJ, Salafranca MN, Adhikari S, Thompson DA, Botti P, Bacon KB, Feng L (1998) Role for neuronally derived fractalkine in mediating interactions between neurons and CX3CR1-expressing microglia. Proc Natl Acad Sci USA 95:10896–10901

Hetier E, Ayala J, Denèfle P, Bousseau A, Rouget P, Mallat M, Prochiantz A (1988) Brain macrophages synthesize interleukin-1 and interleukin-1 mRNAs in vitro. J Neurosci Res 21:391–397

Hickey WF, Kimura H (1988) Perivascular microglial cells of the CNS are bone marrow-derived and present antigen in vivo. Science 239:290–292

Hirota H, Kiyama H, Kishimoto T, Taga T (1996) Accelerated nerve regeneration in mice by upregulated expression of interleukin (IL) 6 and IL-6 receptors after trauma. J Exp Med 183:2627–2634

Hoek RM, Ruuls SR, Murphy CA, Wright GJ, Goddard R, Zurawski SM, Blom B, Homola ME, Streit WJ, Brown MH, Barclay AN, Sedgwick JD (2000) Down-regulation of the macrophage lineage through interaction with OX2 (CD200). Science 290:1768–1771

Jung S, Aliberti J, Graemmel P, Sunshine MJ, Kreutzberg GW, Sher A, Littman DR (2000) Analysis of fractalkine receptor CX(3)CR1 function by targeted deletion and green fluorescent protein reporter gene insertion. Mol Cell Biol 20:4106–4114

Kiefer R, Lindholm D, Kreutzberg GW (1993) Interleukin-6 and transforming growth factor-β1 mRNAs are induced in rat facial nucleus following motoneuron axotomy. Eur J Neurosci 5:775–781

Klein MA, Muller JC, Jones LL, Bluethmann H, Kreutzberg GW, Raivich G (1997) Impaired neuroglial activation in interleukin-6 deficient mice. Glia 19:227–233

Kreutzberg GW (1996) Microglia: a sensor for pathological events in the CNS. Trends Neurosci 19:312–318

Lemke R, Hartig W, Rossner S, Bigl V, Schliebs R (1998) Interleukin-6 is not expressed in activated microglia and in reactive astrocytes in response to lesion of rat basal forebrain cholinergic system as demonstrated by combined in situ hybridization and immunochemistry. J Neurosci Res 51:223–236

März P, Cheng JG, Gadient RA, Patterson PH, Stoyan T, Otten U, Rose-John S (1998) Sympathetic neurons can produce and respond to interleukin 6. Proc Natl Acad Sci USA 95:3251–3256

McCann MJ, O'Callaghan JP, Martin PM, Bertram T, Streit WJ (1996) Differential activation of microglia and astrocytes following trimethyltin-induced brain injury. Neuroscience 72:273–281

Morioka T, Kalehua AN, Streit WJ (1991) The microglial reaction in the rat dorsal hippocampus following transient forebrain ischemia. J Cereb Blood Flow Metab 11:966–973

Morioka T, Baba T, Black KL, Streit WJ (1992) Response of microglial cells to experimental rat glioma. Glia 6:75–79

Müller T, Hanisch UK, Ransom BR (2000) Thrombin-induced activation of cultured rodent microglia. J Neurochem 75:1539–1547

Murphy PG, Grondin J, Altares M, Richardson PM (1995) Induction of interleukin-6 in axotomized sensory neurons. J Neurosci 15:5130–5138

Murphy PG, Borthwick LS, Johnston RS, Kuchel G, Richardson PM (1999) Nature of the retrograde signal from injured nerves that induces interleukin-6 mRNA in neurons. J Neurosci 19:3791–3800

Nishiyori A, Minami M, Ohtani Y, Takami S, Yamamoto J, Kawaguchi N, Kume T, Akaike A, Satoh M (1998) Localization of fractalkine and CX_3CR1 mRNAs in rat brain: does fractalkine play a role in signaling from neuron to microglia? FEBS Lett 429:167–172

Pearson VL, Rothwell NJ, Toulmond S (1999) Excitotoxic brain damage in the rat induces interleukin-1β protein in microglia and astrocytes: correlation with the progression of cell death. Glia 25:311–323

Perry VH, Brown MC, Gordon S (1987) The macrophage response to central and peripheral nerve injury. A possible role for macrophages in regeneration. J Exp Med 165:1218–1223

Popovich PG, Stokes BT, Whitacre CC (1996) Concept of autoimmunity following spinal cord injury: possible roles for T lymphocytes in the traumatized central nervous system. J Neurosci Res 45:349–363

Raivich G, Haas S, Werner A, Klein MA, Kloss C, Kreutzberg GW (1998) Regulation of MCSF receptors on microglia in the normal and injured mouse central nervous system: a quantitative immunofluorescence study using confocal laser microscopy. J Comp Neurol 395:342–358

Schenk D, Barbour R, Dunn W, Gordon G, Grajeda H, Guido T, Hu K, Huang J, Johnson-Wood K, Khan K, Kholodenko D, Lee M, Liao Z, Lieberburg I, Motter R, Mutter L, Soriano F, Shopp G, Vasquez N, Vandevert C, Walker S, Wogulis M, Yednock T, Games D, Seubert P (1999) Immunization with amyloid-β attenuates Alzheimer-disease-like pathology in the PDAPP mouse. Nature 400:173–177

Schwaeble WJ, Stover CM, Schall TJ, Dairaghi DJ, Trinder PKE, Linington C, Iglesias A, Schubart A, Lynch NJ, Weihe E, Schafer MK (1998) Neuronal expression of fractalkine in the presence and absence of inflammation. FEBS Lett 439:203–207

Streit WJ (2002) Physiology and pathophysiology of microglial cell function. In: Streit WJ (ed) Microglia in the regenerating and degenerating CNS. Springer-Verlag, New York, pp 1–14

Streit WJ, Kreutzberg GW (1988) The response of endogenous glial cells to motor neuron degeneration induced by toxic ricin. J Comp Neurol 268:248–263

Streit WJ, Graeber MB, Kreutzberg GW (1988) Functional plasticity of microglia: a review. Glia 1:301–307

Streit WJ, Graeber MB, Kreutzberg GW (1989) Expression of Ia antigens on perivascular and microglial cells after sublethal and lethal neuronal injury. Exp Neurol 105:115–126

Streit WJ, Morioka T, Kalehua AN (1992) MK-801 prevents microglial reaction in rat hippocampus after forebrain ischemia. NeuroReport 3:146–148

Streit WJ, Semple-Rowland SL, Hurley SD, Miller RC, Popovich PG, Stokes BT (1998) Cytokine mRNA profiles in contused spinal cord and axotomized facial nucleus suggest a beneficial role for inflammation and gliosis. Exp Neurol 152:74–87

Streit WJ, Hurley SD, McGraw TS, Semple-Rowland SL (2000) Comparative evaluation of cytokine profiles and reactive gliosis supports a critical role for interleukin-6 in neuron-glia signaling during regeneration. J Neurosci Res 61:10–20

Streit WJ, Conde JR, Harrison JK (2001) Chemokines and Alzheimer's disease. Neurobiol Aging 22:909–913

Tseng GF, Wang YJ, Lai QC (1996) Perineuronal microglial reactivity following proximal and distal axotomy of rat rubrospinal neurons. Brain Res 715:32–43

Ulvestad E, Williams K, Matre R, Nyland H, Olivier A, Antel J (1994) Fc receptors for IgG on cultured human microglia mediate cytotoxicity and phagocytosis of antibody-coated targets. J Neuropathol Exp Neurol 53:27–36

Zhong J, Dietzel ID, Wahle P, Kopf M, Heumann R (1999) Sensory impairments and delayed regeneration of sensory axons in interleukin-6-deficient mice. J Neurosci 19:4305–4313

Zujovic V, Benavides J, Vige X, Carter C, Taupin V (2000) Fractalkine modulates TNF-alpha secretion and neurotoxicity induced by microglial activation. Glia 29:305–315

3 Elucidating Molecular Mechanisms of Alzheimer's Disease in Microglial Cultures

J. Rogers, L.-F. Lue, D.G. Walker, S.D. Yan, D. Stern,
R. Strohmeyer, C.J. Kovelowski

3.1 Introduction 25
3.2 Microglial Cultures from Human Postmortem Brain 27
3.3 Modeling Microglial Activation, Chemotaxis, and Phagocytosis in Response to Aβ 30
3.4 Molecular Mechanisms of Microglial Activation, Chemotaxis, and Phagocytosis of Aβ 35
3.5 Using Human Postmortem Microglia Cultures to Investigate Microglial Responses to Antibody-Opsonized Aβ 38
References 39

3.1 Introduction

Particularly in the context of innate inflammatory mechanisms, microglia appear to play important roles in a wide range of neurodegenerative diseases, including multiple sclerosis, Parkinson's disease, and human immunodeficiency virus (HIV)-associated dementia (reviewed in Banati et al. 1993; Dickson et al. 1993; McGeer et al. 1993). It should not be surprising, then, that microglial activation has been found to be a crucial event mediating inflammatory responses in Alzheimer's disease (AD) (reviewed in Neuroinflammation Working Group 2000).

Ontogenetically, microglia are closely related to peripheral macrophages, and have many functional parallels to macrophages when viewed in the context of central nervous system inflammation (Hulette et al. 1992; Banati et al. 1993; Ling and Wong 1993; Ulvestad et al. 1994a,b). Activated microglia are characterized by their capabilities for differentiation, migration, and phagocytosis (Cotter et al. 1999; Gehrmann et al. 1993; Giulian and Baker 1986), as well as their ability to act as secretory cells that release a wide variety of factors such as cytokines, chemokines, proteolytic enzymes, reactive oxygen species, and complement proteins (reviewed in Neuroinflammation Working Group 2000). Some of these factors may further act on astrocytes, neurons, other microglia, or all these cell types to perpetuate ongoing inflammation (Peterson et al. 1997; Turrin and Plata-Salaman 2000). As the main resident immune effector cells of the nervous system, microglia also express cell-surface molecules required for interacting with different types of immune cells. These molecules include the major histocompatibility complex (MHC) class I and II cell surface glycoproteins for antigen presentation, and immunoglobulin (IgG) Fc receptors (FcγR) that enhance the phagocytosis of antibody-opsonized antigens (Rogers et al. 1988; Gehrmann et al. 1993; Ulvestad et al. 1994a,b; Frautschy et al. 1998). Both classes of molecules are upregulated in AD postmortem brain samples (Rogers et al. 1988; Peress et al. 1993).

Recent studies using an amyloid β peptide (Aβ) immunization approach in Aβ precursor protein (APP) overexpressing transgenic mice (Bard et al. 2000; Schenk et al. 1999) have made understanding microglial interactions with Aβ an especially salient and urgent undertaking. APP transgenic mice immunized with Aβ to produce specific circulating antibodies or passively immunized with Aβ antibodies show increased clearance of cortical and limbic Aβ deposits compared with age-matched controls that have not been immunized. This effect has been attributed to a functional FcR-mediated phagocytic mechanism by MHC II immunoreactive, activated microglia (Schenk et al. 1999; Bard et al. 2000), and has suggested two contrasting outcomes of microglial interactions with Aβ. On the one hand, microglial phagocytosis of opsonized Aβ should be beneficial in and of itself and, if succesful, should also remove a major stimulus for AD neuroinflammation, interactions of microglia and inflammatory mediators with Aβ (reviewed in Neuroinflammation Working Group 2000). Alternatively, activation of microglial Fc receptors could engender a wide range of destructive pro-inflammatory mechanisms such as the respiratory burst that lead to the well known process of antibody-dependent cell-mediated cytotoxicity

(ADCC) (Ulvestad et al. 1994a,b; Edwards et al. 1997). Because immunization therapy for AD is now well into phase I clinical trials, it has become important to evaluate mechanistically how microglia interact with Aβ and opsonized Aβ, and to understand the consequences of such interactions on the secretion of cytokines and other inflammatory molecules.

3.2 Microglial Cultures from Human Postmortem Brain

In view of the pivotal role that microglial activation plays in AD pathogenesis, we have developed human microglial cultures from rapid autopsies of AD and non-demented elderly (ND) brains to model molecular mechanisms of microglial activation [Walker et al 1995; Lue et al. 1996b, in press (a,b)]. We believe these cultures provide an optimal AD research preparation in three significant ways: they are appropriate with respect to developmental stage, species, and disease state; they offer opportunities to study dynamic responses in living cells (an impossibility with preserved postmortem samples); and they are highly amenable to research into specific molecular mechanisms.

Over the last 6 years, we have isolated microglia from rapid autopsies of 223 AD and ND patients enrolled in the Sun Health Research Institute Brain Autopsy program. The brain samples were collected with informed, premortem consent and had an average postmortem delay of 3.11±0.55 h [Lue et al. 1996b, in press (a)]. The isolation procedure is one modified from methods developed originally by Kim for the isolation of oligodendrocytes (Kim et al. 1983). Typically, 20–40 g of cortical tissue from the superior and middle frontal gyri of the right hemisphere, and 8–10 g of corpus callosum at the level of the genu are taken at autopsy. The cortical samples usually contain 60%–80% gray matter. The finely minced brain tissues are dissociated by digestion with trypsin at a final concentration of 0.15% for 45 min with agitation in a 37°C water bath. Dissociated brain cells are then subjected to washing, filtering, and centrifugation through Percoll gradients to separate viable cells from erythrocytes and myelin debris. For each case, a total viable cell count is determined by trypan blue exclusion before the final resuspension of cells in Dulbecco's modified Eagle's medium, supplemented with 10–50 μg/ml gentamicin sulfate and 10% fetal bovine serum.

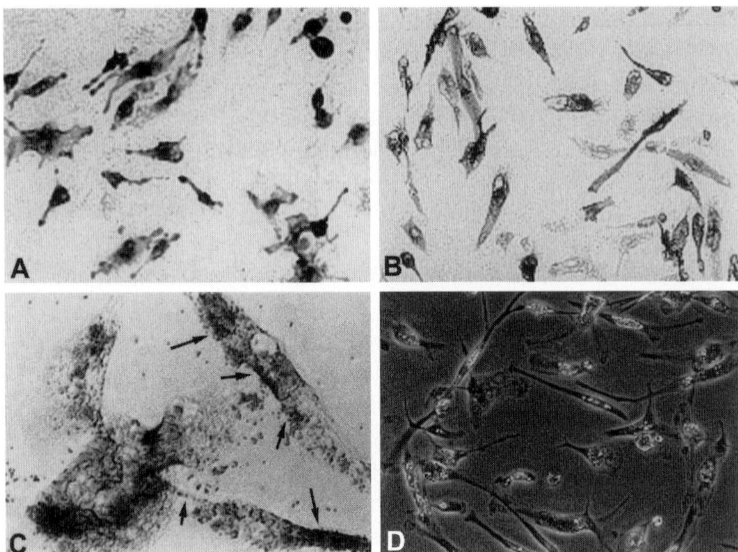

Fig. 1A–D. Characterization of AD and ND microglia in vitro. **A** CD68 immunoreactivity. **B** MHCII (LN3) immunoreactivity. **C** Phagocytosis of latex beads, which appear intracellularly as clusters (*arrows*) and as individual beads. **D** Phase contrast 1 week post-plating

At first plating, the cultures initially contain a mixture of different glial types. We have found, however, that each glial type has different adherence properties that can be used to advantage to isolate one class from another. That is, postmortem AD and ND microglia are the first to adhere to plastic culture surfaces, which is also the case for human adult microglia isolated from biopsy brain samples, but not for human fetal microglia (Kim et al. 1983; Lee et al. 1992; Williams et al. 1992). Highly pure microglial cultures (>99%) can therefore be obtained simply by pouring off the non-adherent cells from non-coated culture plates within 24 h of initial plating. The non-adherent cells are then replated on poly-L-lysine coated plates to isolate astrocytes and/or oligodendrocytes (Lue et al. 1996b).

Routine purity assessments of the microglial cultures show them to be consistently positive for CD68, MHC II, HLA-DR, and low-density lipoprotein receptors (Fig. 1A,B), and to be consistently negative for glial fibrillary acidic protein, galactocerebroside, neurofilaments, and other non-microglia markers. The cultured microglia are phagocytic (Fig. 1C), whereas our cultured astrocytes are not. Viewed with phase-contrast optics (Fig. 1D), AD and ND microglia in vitro are morphologically distinct from AD and ND astrocytes and oligodendrocytes. Their dark and granular cell bodies are usually surrounded by bright phase and contain cytoplasmic vacuoles. In the first 48 h in culture, the microglia tend to be small in size and round, oval, or slightly bipolar in shape. Within 1 week, their cell bodies tend to enlarge, and more heterogeneous morphologies emerge.

If replating for specific experiments is required, the isolated microglia can be readily detached at 12–14 days in vitro using 0.25% trypsin at 37°C or 2 mM ethylenediaminetetraacetate (EDTA) in phosphate-buffered saline on ice. The latter procedure yields much better preservation of cell surface receptor molecules. Although the microglia can be detached and replated at other intervals in vitro, we find the cells to be optimally viable and phenotypically most similar to AD and ND microglia in situ (e.g., in terms of the percentage of MHC II immunoreactive cells) at approximately 2 weeks post-plating. For these reasons, almost all of our experiments with the cells have been conducted at this time. With prolonged periods of culture, microglial adherence and activation increase, causing difficulties removing the cells for replating, and more aggressively activated, amoeboid morphologies emerge.

At first plating, total yields (all glial types) have averaged 2.78 million cells/gram brain tissue. After isolation and 14 days in vitro, microglial yields have averaged 0.15 million cells/gram brain tissue. Although we have not observed a significant difference in total glial yields or microglial yields when AD cases are compared to ND cases, there is a trend to lower yields from AD samples.

Several other groups have endeavored to produce isolated glial cultures from human adults. Kim et al. (1983) and Lisak et al. (1981) originally used Percoll gradient centrifugation and the differential adherence properties of cells to develop enriched oligodendrocyte cultures from adult animal and human white matter. Later, a similar procedure was used to prepare microglia and astrocytes from biopsy brain samples of human adults and postmortem brains (Williams et al. 1992; Whittemore et al. 1993). DeGroot and colleagues (2000) have also followed similar principles to isolate microglia, astrocytes, and oligodendrocytes from postmortem nondemented and demented patients. In their protocol, percoll centrifugation is performed at a lower speed for a shorter time compared to our protocol, which employs high-speed centrifugation (13,000–15,000 rpm) for 30–40 min at 4°C. Microglial yields from these procedures reportedly range from 0.017 to 0.1 million cells/gram of cortical gray matter, and from 0.1 to 0.2 million cells/gram of white matter, slightly lower than our yields from AD and ND autopsies. Microglia isolated by DeGroot and colleagues (2000) appear to exhibit similar phenotypic and functional characteristics as microglia recovered by our protocol, including transitions to a more ameboid morphology with time in culture. In addition, these investigators have demonstrated a 10%–20% proliferation in response to granulocyte-macrophage-colony stimulating factor (GM-CSF) treatment, a finding consistent with previous observations of GM-CSF-stimulated proliferation in young adult (but not aged) rodent microglial cultures (Rozovsky et al. 1998), as well as human fetal microglial cultures (Lee et al. 1992). By contrast, independent observations from our laboratory and others have indicated that in the absence of exogenous growth factors, adult microglia usually do not proliferate in vitro (Williams et al. 1992), nor do cultured fetal microglia if they are separated from astrocytes and remain unexposed to exogenous growth factors (Lee et al. 1992).

In summary, our research [Lue et al. 1996b, in press (a,b); Walker et al. 1995] and that of others (Lisak et al. 1981; Williams et al. 1992; Kim et al. 1993; Whittemore et al. 1993; DeGroot et al. 2000) has demonstrated that it is possible to reproducibly isolate and maintain highly purified cultures of microglia from postmortem adults. These cultures not only provide new experimental opportunities for studying human neurologic disorders, but also circumvent potential ethical and political dilemmas that, rightly or wrongly, have sometimes obtained with cultures from aborted human fetuses.

3.3 Modeling Microglial Activation, Chemotaxis, and Phagocytosis in Response to Aβ

Two of the most distinctive features of microglia in the AD brain are their high degree of activation, as evidenced by their elevated expression of MHC II, and the close association of MHC II immunoreactive micro-

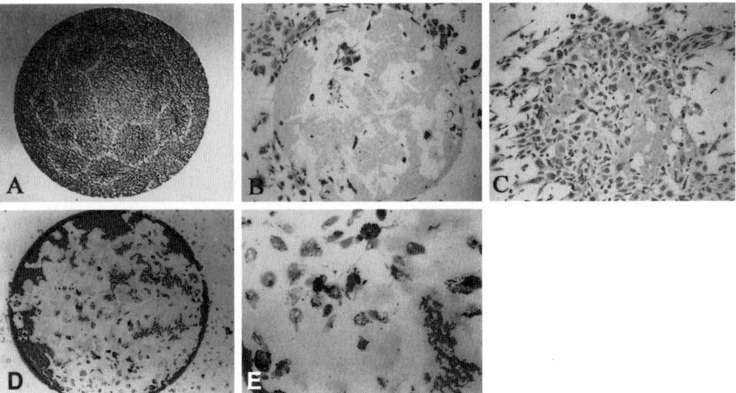

Fig. 2A–E. Microglial responses to Aβ in vitro. **A** Aβ spot pre-plating. **B** Aβ spot approximately 2 weeks post-plating with AD microglia. **C** Aβ spot 3 weeks post-plating with AD microglia. **D** Phagocytosis of Aβ spot. **E** As the Aβ spot is cleared, the microglia become intensely Aβ immunoreactive

glia with mature Aβ plaques and neurofibrillary tangles (Rogers et al. 1988; Perlmutter et al. 1990; Lue et al. 1996a). The graded concentrations of MHC-immunoreactive microglia in the vicinity of Aβ deposits in AD cortex (Rogers et al. 1988) and in APP_{SW} transgenic mouse cortex (Frautschy et al. 1998) strongly suggest that Aβ may stimulate both microglial activation and chemotaxis. Likewise, ultrastructural and other studies of the AD brain have suggested microglial phagocytosis of Aβ (reviewed in Rogers and Lue 2001).

We have modeled microglial activation, chemotaxis, and phagocytosis in response to Aβ plaques by preparing culture plates with deposits of aggregated Aβ [Lue et al., in press (a,b)], similar to a previously described system (Canning et al. 1993; Wujek et al. 1996). Aβ1-42 deposits (pre-aggregated at 0.5 or 1 mM at 37°C for 2 h) are formed as spots and air-dried on poly-L-lysine coated culture plate surfaces. The spots exhibit the compact β-pleated sheet conformation of amyloid and are thioflavin S positive. Primary microglial cultures are added to the wells. At this initial plating, small ramified microglia are present, along with a mixture of other glial cell types. Twenty-four hours later, non-adherent cells are removed to provide relatively pure (>99%) microglial cultures. Over the next few days, the microglia become more vacuo-

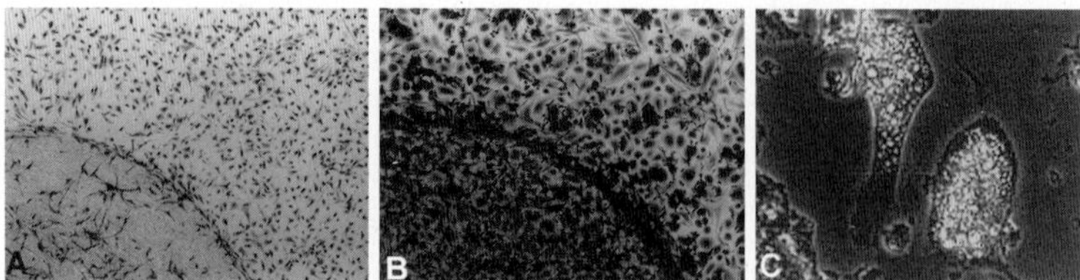

Fig. 3. A–C. With progressive exposure to Aβ spots in vitro, AD and ND microglia become more intensely MHC II immunoreactive (**A**, 1 week post-plating; **B**, 3 weeks post-plating), and display highly vacuolated, ameboid morphologies (**C**)

lated, and appear to migrate toward the Aβ spots (Fig. 2A–C), initially aligning around their perimeter and ultimately covering much of their surface. Aggressive phagocytosis ensues (Fig. 2D) and the microglia become intensely immunoreactive for Aβ (Fig. 2E). Heightened MHC II expression is evident (Fig. 3A,B). By 3 weeks the microglia are filled with cytoplasmic vacuoles and display a highly activated, ameboid morphology (Fig. 3C).

Similar findings have been reported for murine peritoneal macrophages and rat microglia (Davis et al. 1992; Meda et al. 1995b). Both Aβ1-42 and Aβ25-35, but not scrambled Aβ1-42, effectively induce cellular migration (Davis et al. 1992; Nakai et al. 1998). Likewise, when plated on AD cortical tissue sections, rodent microglia migrate to Aβ plaques on the sections (Ard et al. 1996; Bard et al. 2000), an effect that is enhanced if the sections are pretreated with Aβ antibodies (Bard et al. 2000). The chemotactic properties of Aβ can also be demonstrated with peripheral monocytes, which migrate across a human blood–brain barrier model when Aβ is present on the other side (Fiala et al. 1998).

In addition to modeling microglial chemotaxis and phagocytosis, our culture preparations provide opportunities to assess the secretory properties of postmortem brain-derived human microglia under basal and Aβ-stimulated conditions. We have summarized our published (Walker et al. 1995) and unpublished [Lue et al., in press (a,b)] data in Table 1 to show that human microglia in culture constitutively express a wide variety of AD-related proteins and mRNAs, and increase the expression of many of these factors when exposed to Aβ. These proteins and mRNAs are not only well studied in AD (reviewed in Neuroinflamma-

Table 1. Expression of inflammatory factors by AD and ND microglia in vitro

Classification	Molecules (other names)	Detection	Constitutive	Aβ-enhanced
CD antigens	CD11b (2M intergrin)	ICC	+	
	CD32 (FcRII)	ICC	+	
	CD 35 (CR1)	RT-PCR	+	
	CD40	ICC, RT-PCR, WB	+	
	CD45 (Leucocyte common antigen)	ICC	+	
	CD54 (ICAM-1)	RT-PCR	+	
	CD64 (FcRI)	ICC	+	
	CD68 (macrosialin)	ICC	+	+
	CD87 (uPA receptor)	ICC, RT-PCR, WB	+	
	CD91 (α2 macroglobulin receptor)	RT-PCR	+	
	CD115 (M-CSF receptor)	RT-PCR	+	
	CD154 (CD40 ligand)	ICC, RT-PCR, WB	+	
Antigenic recognition	MHC II	ICC, WB	+	+
Complement proteins	C1q	ICC, ELISA, RT-PCR	+	
	C 3	RT-PCR	+	
Complement inhibitors	C1 NH	RT-PCR	+	
	C4 BP	ICC, RT-PCR	+	
Cytokines, chemokines, growth factors, and receptors	IL-1α	RT-PCR	+	
	IL-1β	ELISA, RT-PCR	+	+
	IL-1rα	RT-PCR	+	
	IL-6	ELISA, RT-PCR	+	+
	IL-8	ELISA	+	+
	MCP-1	ELISA	+	+
	M-CSF	ICC,ELISA	+	+
	M-CSF R(c-fms)	RT-PCR	+	
	MIP-1α	ELISA	+	+
	TNF-α	ELISA, RT-PCR	+	+
	TGF-β	RT-PCR	+	
	TNF-αRI, I	RT-PCR	+	
	TNF-αRI, II	RT-PCR	+	
Aβ-binding receptors	RAGE	ICC, RT-PCR, WB	+	+
	MSR	ICC, RT-PCR, WB	+	
	MPO	RT-PCR	+	
Extracellular matrix component	Perlecan	RT-PCR	+	
Membrane associated protein	APP	ICC, RT-PCR	+	

APP, amyloid precursor protein; C1 NH, C1 inhibitor; C4 BP, C4-binding protein; ICAM-1, intercellular adhesion molecules; ICC, immunocytochemistry; ELISA, enzyme-linked immunoassay; IL, interleukin, FcR, immunoglobulin Fc receptor; MCP, monocyte chemotactic peptide; M-CSF, macrophage colony stimulating factor; MHC, major histocompatibility complex; MIP-1, macrophage inflammatory protein; MPO, myeloperoxidase; MSR, macrophage scavenger receptor; RT-PCR, reverse transcription polymerase chain reaction; TGF, tissue growth factor; TNF, tumor necrosis factor; RAGE, receptor for advanced glycation end products; uPA, urokinase plasminogen activator.

tion Working Group 2000), but are functionally linked to a broad spectrum of inflammatory mechanisms, including antigen recognition (MHC II), inflammatory responses (CD40, CD40L, complement proteins and inhibitors, cytokines, chemokines, growth factors, and their receptors), phagocytosis [CD11b, CD32, CD64, CD68, macrophage scavenger receptor (MSR)], adhesion (CD54), proteolysis [urokinase plasminogen activator (uPA), urokinase plasminogen activator receptor (uPAR)], and oxidative damage [myeloperoxidase (MPO)]. In addition, our cultured microglia produce peptides that have been directly implicated in AD pathophysiology, including Aβ-binding proteins [MSR, C1q, receptor for advanced glycation end products (RAGE)] and mRNA and protein for APP. The presence of these inflammatory and AD-related markers in AD brain and culture models has been previously reviewed in detail (Neuroinflammation Working Group 2000).

With regard to constitutive expression, microglial secretion of most of the inflammatory factors we have studied so far does not differ in AD and ND cultures. Two exceptions are C1q, the first classical pathway complement component, and the growth factor M-CSF. C1q is of special interest because it is known to be present from the very earliest stages of plaque development in Down's syndrome (Stoltzner et al. 2000), has potent enhancing effects on Aβ aggregation (Webster and Rogers 1996; Webster et al. 1997a,b), is activated by Aβ to initiate the full complement cascade (Rogers et al. 1992; Webster et al. 1997c), and is not only elevated in the AD cortex (Brachova et al. 1993; Walker et al. 1995), but is elevated in a manner disproportionate to its primary regulator, C1 inhibitor, which is not increased in AD cortex (Veerhuis et al. 1999). M-CSF also has substantial relevance to AD pathophysiology because of its potential role in microglial activation and proliferation, and because of its interactions with RAGE and Aβ (see below). If a truly constitutive, elevated expression of C1q and M-CSF extends to microglia in situ in the elderly brain, then such a condition might well predispose individuals possessing such traits to the inflammatory sequelae of AD.

In addition to constitutive expression, the vast majority of inflammatory factors measured in our cultured microglia show significant increases after exposure to doses of Aβ as low as 10 nM [Lue et al., in press (a)]. The findings include elevated mRNA for IL-1β, and elevated proteins for pro-IL-1β, IL-6, IL-8, TNF-α, macrophage inflammatory protein (MIP)-1α, monocyte chemotactic peptide (MCP)-1, and M-CSF. These results are consistent with other documented effects of Aβ-stimulation on cytokine secretion from rodent microglia, differentiated or non-differentiated human monocytes, murine microglial cell lines, and fetal human microglia in vitro, as well as elevated concentrations of these factors in the AD brain (reviewed in Neuroinflammation Working Group 2000). The in vitro preparations, however, provide opportunities to examine many variables that could not be studied with postmortem tissue, such as the effects of Aβ secondary and tertiary structure or the effects of priming agents. Meda and colleagues (1995a,b, 1996, 1999), for example, have investigated the induction of a range of cytokines by the Aβ25-35 fragment using rodent microglia and human monocyte cultures. Exposure of the cells to Aβ25-35 alone induced expression of IL-8, IL-1β, IL-1RA, MCP-1, and MIP-1α. After co-stimulation with interferon-γ, Aβ25-35 also potentiated the induction of NO and TNF-α .

3.4 Molecular Mechanisms of Microglial Activation, Chemotaxis, and Phagocytosis of Aβ

The molecular mechanisms that underlie microglial activation, chemotaxis, and phagocytosis in response to Aβ are only just beginning to be understood. The fact that these processes occur at all, however, strongly suggests that some sort of ligand-receptor interaction must be taking place. Multiple candidates are available, including RAGE, the MSR, the formyl peptide receptor, and others.

Microglia can bind aggregated Aβ via RAGE (Yan et al. 1997, 2000). These findings are consistent with other data implicating RAGE/Aβ binding in a wide range of microglial behaviors. Previous observations in the BV-2 murine microglial cell line and primary rodent microglial cultures, for example, have indicated that cell surface RAGE binding to Aβ helps mediate chemotaxis and haptotaxis (Yan et al. 1997). Blockade of RAGE with anti-RAGE $F(ab')_2$ has been shown to reduce Aβ-induced microglial directional migration toward Aβ, to prevent BV-2 cells from being bound to membranes pre-adsorbed with Aβ1-40, and to permit microglial migration towards the chemoattractant formylmethionine-leucine phenylalanine (fMLP) (Yan et al. 1997, 1999, 2000).

Complex interactions of RAGE with M-CSF are likely to be involved in any RAGE-mediated microglial responses. RAGE immunoreactivity expressed at the cell surface of human postmortem microglia can be increased with 1 μM aggregated Aβ1-42 and 50 ng/ml M-CSF [Lue et al., in press (b)]. Reverse transcriptase polymerase chain reaction (RT-PCR) assays also demonstrate that RAGE mRNA expression in human microglia can be induced by 5 μM Aβ stimulation for 24 h. In addition, when microglial cultures are pretreated with 10 μg/ml anti-RAGE $F(ab')_2$, constitutive M-CSF secretion is decreased more than 90%, and Aβ-induced M-CSF secretion is inhibited by 67.5% [Lue et al., in press (b); Yan et al. 1997].

We have also exploited our chemotaxis model to investigate the influence of RAGE/M-CSF/Aβ interactions. Microglial chemotaxis to and accumulation on Aβ spots is significantly reduced when the cultures are treated with 10 μg/ml of anti-RAGE $F(ab')_2$. By contrast, RAGE mRNA is upregulated in our chemotaxis model after treatment with 50 ng/ml M-CSF for 24 h, an effect that may occur in an autocrine fashion, since we have further demonstrated that our microglial cultures express mRNA for the M-CSF receptor (c-fms) [Lue et al., in press (b)]. Collectively, then, these results suggest that in an Aβ-enriched environment such as the AD cortex an increase in M-CSF secretion resulting from Aβ-RAGE binding could accentuate microglial activation and chemotaxis by a positive feedback loop. That is, RAGE/Aβ binding increases the expression of M-CSF, which increases the expression of RAGE. Heightened RAGE expression, in turn, would foster more Aβ binding, more M-CSF production, and more RAGE expression. All these processes should result in enhanced microglial activation and chemotaxis in the vicinity of Aβ deposits.

Despite the collective findings on RAGE, it would be a mistake to ignore the likely contributions of other Aβ-receptive mechanisms. For example, the fMLP (major late promoter) receptor, fMLPR, has been shown to bind Aβ causing subsequent activation of monocyte/microglia cells (Lorton 1997; Lorton et al. 2000). In contrast to RAGE, binding of Aβ to fMLPR activates a calcium-dependent, G-protein-linked second messenger system similar to chemokine receptor activation (Lorton 1997). A third obvious candidate for mediating Aβ/microglial interactions is the MSR, which is expressed by microglia co-localized with Aβ deposits in the AD brain (El Khoury et al. 1996). In our culture chemotaxis model, increased MSR-A immunoreactivity (using a mouse anti-human MSR-A antibody generously provided by Drs. T. Kodoma and M. Honda of the University of Tokyo) is detected on microglia recruited to Aβ spots (L.F. Lue et al., unpublished data). A fourth candidate Aβ receptor – one whose identity is still being characterized – has been reported by Giulian and colleagues (Giulian et al. 1998).

In addition to receptor mechanisms, several other factors are likely to contribute to the activation and migration of microglia toward Aβ deposits, particularly soluble mediators secreted by Aβ-activated microglia. For example, M-CSF, which is increased after the exposure of microglia [Lue et al., in press (a)] and neurons (Yan et al. 1999) to Aβ, induces microglial chemotaxis, proliferation, increased MSR expression, and enhanced microglial survival (Giulian and Baker 1986; Tomozawa et al. 1996). Aβ-stimulated secretion of M-CSF could therefore contribute substantially to the accumulation of activated microglia in and around Aβ plaques. Notably, when Aβ plaque cores from AD patients are injected into rat brain, proliferation of microglia occurs (Frautschy et al. 1992), consistent with the effects of cell growth factors such as M-CSF. Furthermore, some chemokines and chemokine receptors belonging to the CCR2, CCR3, and CCR5 families appear to be upregulated in activated microglia associated with amyloid deposits (Ishizuka et al. 1997; Xia and Hyman 1999). Among these, MCP-1, and MIP-1α have been shown to be induced by Aβ1-42 treatment in our human postmortem cultures [Lue et al. in press (a)]. Since these chemokines are known to stimulate chemotaxis, their upregulation by microglia at sites of Aβ deposition would be expected to provide a gradient for chemotaxis to such sites.

The multiplicity of potential mechanisms by which microglial responsiveness to Aβ may occur should provide an important caveat to knockout and other results that have been taken to suggest that RAGE, the MSR, or some other candidate mechanism is not involved in microglial/Aβ interactions (Huang et al. 1999). Clearly, if multiple mechanisms can do the job, then knocking out one or even two of the candidate mechanisms might still permit fairly rigorous microglial responses. Thus, the absence of an effect on a microglial response to Aβ after knockout of a putative Aβ receptor such as RAGE should not necessarily be taken as evidence that the receptor plays no role in the response.

3.5 Using Human Postmortem Microglia Cultures to Investigate Microglial Responses to Antibody-Opsonized Aβ

Persistence of Aβ plaques in AD brains may be caused by Aβ overproduction, poor degradation, and/or failure of phagocytosis by microglia. In vitro experiments have shown that although cultured microglia can phagocytose Aβ, they have limited ability to degrade it compared to other proteins (Paresce et al. 1997; Bard et al. 2000). By contrast, immunization of PD-APP transgenic mice with Aβ peptides or Aβ opsonization in vitro dramatically enhances the clearance of Aβ deposits by activated microglia (Schenk et al. 1999; Bard et al. 2000).

Fc-receptor-mediated phagocytosis is normally accompanied by the respiratory burst and other potentially cytotoxic responses of scavenger cells (Ulvestad et al. 1994a,b; Edwards et al. 1997). It is therefore somewhat surprising that immunized PD-APP mice do not evidence detectable cytotoxicity and, in fact, show decreased behavioral deficits compared to non-immunized controls (Janus et al. 2000; Morgan et al. 2000). Why this should be remains unclear. One obvious possibility is that the innate inflammatory responses of mice to human Aβ are blunted compared to those of humans. Complement activation by human Aβ, for example, is significantly less for mouse complement than human complement (Webster et al. 1999). Moreover, the time frame for potential inflammatory toxicity in transgenic mice is over a period of months, whereas it is as much as a decade for human AD patients. Differences in the physical chemical properties of Aβ deposits in transgenic APP-overexpressing mice and human AD patients may also be a factor. Kuo and co-workers (2001), for example, have shown that Aβ deposits in APP23 transgenic mice are dramatically more soluble than those in AD cortex, perhaps making them much easier to clear. If so, then the Fc-receptor-mediated clearance of Aβ in transgenic mice might require only modest Aβ opsonization that is below the threshold for stimulation of significant cytotoxic, secondary inflammatory actions.

Two other mechanisms may be worth pursuing in our efforts to understand the apparent absence of cytotoxicity in Aβ-immunized transgenic mice. The respiratory burst induced by challenge of macrophages with IgG-coated erythrocytes has been reported to be depressed by the concomitant release of arachidonic acid (Raley et al. 1999), and Fc-mediated phagocytosis has been shown to stimulate synthesis of the anti-inflammatory cytokine IL-10 (Sutterwala et al. 1998). It is possible, therefore, that Fc-mediated phagocytosis of Aβ by microglia taps into pathways that favor the production of immunosuppressive cytokines rather than the usual mix of pro- and anti-inflammatory acute phase reactants. Much research is needed to address such speculations. Microglial cultures that are appropriate with respect to species, developmental stage, and disease state should help provide answers at the cellular and molecular level.

Acknowledgements. We thank Dr. Thomas Beach, Lucia Sue, Sarah Scott, and Kathryn Layne for their well-organized autopsy program, and Kyle Mueller, Sonal Vallabhbhai, Rajalakshmi Seetharaman, Kelly Anderson, Miranda Nowlin, and Julia Nimlos for technical assistance. This work was supported by NIA AGO7367, as well as grants from the Alzheimer's Association and the Arizona Alzheimer's Research Center.

References

Ard MD, Cole GM, Wei J, Mehrie AP, Fratkin JD (1996) Scavenging of Alzheimer's amyloid beta-protein by microglia in culture. J Neurosci Res 43:190–192

Banati RB, Gehrmann J, Schubert P, Kreutzberg GW (1993) Cytotoxicity of microglia. Glia 7:111–118

Bard F, Cannon C, Barbour R, Burke RL, Games L, Grajeda H, Guido T, Hu K, Huang J, Johnson-Wood K, Khan K, Kholodenko D, Lee M, Lieberburg I, Motter R, Nguyen M, Soriano F, Vasquez N, Weiss K, Welch B, Seubert P, Schenk D, Yednock T (2000) Peripherally administered antibodies against amyloid β-peptide enter the central nervous system and reduce pathology in a mouse model of Alzheimer disease. Nat Med 6:916–919

Brachova L, Lue, L-F, Schultz J, El Rashidy T, Rogers J (1993) Association cortex, cerebellum, and serum concentrations of C1Q and factor B in Alzheimer's disease. Mol. Brain Res 18:329–334

Canning DR, Mckeon RJ, DeWitt DA, Perry G, Wujek JR, Frederickson RC, Silver J (1993) beta-Amyloid of Alzheimer's disease induces reactive gliosis that inhibits axonal outgrowth. Exp Neurol 124:289–98

Cotter RL, Burke WJ, Thomas VS, Potter JF, Zheng J, Gendelman HE (1999) Insights into the neurodegenerative process of Alzheimer's disease: a role for mononuclear phagocyte-associated inflammation an d neurotoxicity. J Leukoc Biol 65:416–427

Davis, J.B, McMurray HF, Schubert D (1992) The amyloid beta-protein of Alzheimer's disease is chemotactic for mononuclear phagocytes. Biochem Biophys Res Commun 189:1096–1100

DeGroot CJA, Montagne L, Janssen I, Ravid R, Van Der Valk P, Veerhuis R (2000) Isolation and characterization of adult microglial cells and oligodendrocytes Derived from postmortem human brain tissue. Brain Res Proto 5:585–594

Dickson DW, Lee SC, Mattiace LA, Yen SHC, Brosnan C (1993) Microglia and cytokines in neurological disease, with special reference to AIDS and Alzheimer's disease. Glia 7:75–83

Edwards SW, Watson F, Gasmi L, Moulding DA, Quayle JA (1997) Activation of human neutrophils by soluble immune complexes: role of Fc gamma RII and Fc gamma R III b in stimulation of the respiratory burst and elevation of intracellular Ca^{2+}. Ann NY Acad Sci 832:341–357

El Khoury J, Hickman SE, Thomas CA, Cao L, Silverstein SC, Loike JD (1996) Scavenger receptor-mediated adhesion of microglia to beta-amyloid fibrils. Nature 382:716–719

Fiala M, Zhang L, Gan Z, Sherry B, Taub D, Graves MC, Hama S, Way D, Weinand M, Witte M, Lorton D, Duo Y-M, Roher AE (1998) Amyloid-β induces chemokine secretion and monocyte migration across a human blood-brain barrier model. Mol Med 4:480–489

Frautschy SA, Cole GM, Baird A (1992) Phagocytosis and deposition of vascular beta-amyloid in rat brains injected with Alzheimer beta-amyloid. Am J Pathol 140:1389–1399

Frautschy SA, Fusheng Y, Irrizarry M, Hyman B, Saido TC, Hsiao K, Cole GM (1998) Microglial response to amyloid plaques in APP_{sw} transgenic mice. Am J Path 152:307–17

Gehrmann J, Banati RB, Kreutzberg GW (1993) Microglia in the immune surveillance of the brain: human microglia constitutively express HLA-DR molecules. J Neuroimmunol 48:189–198

Giulian D, Baker TJ (1986) Characterization of amoeboid microglia isolated from developing mammalian brain. J Neurosci 6:21–28

Giulian D, Haverkamp LJ, Yu J, Karshin W, Tom D, Li J, Kazanskaia, Kirkpatrick JB, Roher AE (1998) The HHQK domain of beta-amyloid provides a structural basis for the immunopathology of Alzheimer's disease. J Biol Chem 273:29719–29726

Huang F, Buttini M, Wyss-Coray T, McConlogue L, Kodama T, Pitas RE, Mucke L (1999) Elimination of the class A scavenger receptor does not affect amyloid plaque formation or neurodegeneration in transgenic mice expressing human amyloid protein precursors. Am J Pathol 155:1741–1747

Hulette CM, Downey BT, Burger PC (1992) Macrophage markers in diagnostic neuropathology. Am J Surg Pathol 16:493–499

Ishizuka K, Kimura T, Igata-yi R, Katsuragi S, Takamatsu J, Miyakawa T (1997) Identification of monocyte chemoattractant protein-1 in senile plaques and reactive microglia of Alzheimer's disease. Psych Clin Neurosci 51:135–138

Janus C, Pearson J, McLaurin J, Mathews PM, Jiang Y, Schmidt SD, Chishti MA, Horne P, Heslin D, French J, Mount HT, Nixon RA, Mercken M, Bergeron C, Fraser PE, St George-Hyslop P, Westaway D (2000) Aβ peptide immunization reduces behavioral impairment and plaques in a mouse model of Alzheimer's disease. Nature 408:979–982

Kim Su, Sato Y, Silberberg DH, Pleasure DE, Rorke LB (1983) Long-term culture of human oligodendrocytes. J Neurol Sci 62:295–301

Kuo Y-M, Kokjohn TA, Beach TG, Sue LI, Brune D, Lopez JC, Kalback WM, Abramowski D, Sturchler-Pierrat C, Staufenbiel M, Roher AE (2001) Comparative analysis of amyloid β chemical structure and amyloid plaque morphology of transgenic mouse and Alzheimer's disease brains. J Biol Chem 276:12991–12998

Lee SC, Liu W, Brosnan CF, Dickson DW (1992) Characterization of primary human fetal dissociated central nervous system cultures with an emphasis on microglia. Lab Investig 67:465–476

Ling EA, Wong WC (1993) The origin and nature of ramified and amoeboid microglia: a historical review and current concepts. Glia 7:10–18

Lisak R, Pleasure D, Manning M, Saida T (1981) Long-term culture of bovine oligodendroglia isolated with a Percoll gradient. Brain Res 113:165–170

Lorton D (1997) β-amyloid-induced IL-1β release from an activated human monocyte cell line is calcium- and G-protein-dependant. Mech Aging Dev 94:119–122

Lorton D, Schaller J, Lala A, De Nardin E (2000) Chemotactic-like receptors and Abeta peptide induced responses in Alzheimer's disease. Neurobiol Aging 21:463–473

Lue LF, Brachova L, Civin WH, Rogers J (1996a) Inflammation, AB deposition, and neurofibrillary tangle formation as correlates of Alzheimer's disease neurodegeneration. J Neuropathol Exp Neurol 55:1083–1088

Lue LF, Brachova L, Walker DG, Rogers J (1996b) Characterization of glial cultures from rapid autopsies of Alzheimer's and control patients. Neurobiol Aging 17:421–429

Lue LF, Rydel R, Brigham EF, Yang LB, Hampel H, Murphy Jr. GM, Brachova L,Yan SD, Walker DG, Shen Y, Rogers J (2002a) Inflammatory repertoire of Alzheimer's disease and non-demented elderly microglia in vitro. Glia (in press)

Lue LF, Walker DG, Brachova L, Beach TG, Rogers J. Schmidt AM, Stern D, Yan SD (2002b) Microglial receptor for advanced glycation endproducts

(RAGE) in Alzheimer's disease: identification of cellular activation mechanism. Exp Neurol (in press)

McGeer P, Kawamata T, Walker DG, Akiyama H, Tooyama I, McGeer EG (1993) Microglia in degenerative neurological disease. Glia 7:84–92

Meda L, Bonaiuto C, Szendrei GI, Ceska M, Rossi F, Cassatella MA (1995a) Beta-amyloid (25–35) induces the production of interleukin-8 from human monocytes. J Neuroimmunol 59:29–33

Meda L, Baron P, Prat E, Scarpini E, Scarlato G, Cassatella MA, Rossi F (1999) Proinflammatory profile of cytokine production by human monocytes and murine microglia stimulated with beta-amyloid. J Neuroimmunol 93:4–52

Meda L, Cassatella MA, Szendrei GI, Otvos L, Baron P, Villalba M, Ferrari D, Rossi F (1995b) Activation of microglial cells by beta-amyloid protein and interferon-gamma. Nature 374:647–650

Meda L, Bernaconi S, Bonaiuto C, Sozzani S, Zhou D, Otvos L, Mantovani A, Rossi F, Cassatella MA (1996) Beta-amyloid (25–35) peptide and IFN-gamma synergistically induce the production of the chemotactic cytokine MCP-1/JE in monocytes and microglial cells. J Immunol 157:1213–1218

Morgan D, Diamond DM, Gottschall PE, Ugen KE, Dickey C, Hardy J, Duff K, Jantzen P, DiCarlo G, Wilcock D, Connor K, Hatcher J, Hope C, Gordon M, Arendash GW (2000) Aβ peptide vaccination prevents memory loss in an animal model of Alzheimer's disease. Nature 408:982–985

Nakai M, Hojo K, Taniguchi T, Terashima A, Kawamata T, Hashimoto T, Maeda K, Tanaka C (1998) PKC and tyrosine kinase involvement in amyloid beta (25–35)-induced chemotaxis of microglia. NeuroReport 9:3467–3470

Neuroinflammation Working Group (2000) Inflammation and Alzheimer's disease. Neurobiol Aging 21:383–421

Paresce DM, Chung H, Maxfield FR (1997) Slow degradation of aggregates of the Alzheimer's disease amyloid beta-protein by microglial cells. J Biol Chem 29390–29397

Peress NS, Fleit HB, Perillo E, Kuljis R, Pezzullo C (1993) Identification of FcγRI, II and III on normal human brain ramified microglia and on microglia in senile plaques in Alzheimer's disease. J Neuroimmunol 48:71–80

Perlmutter LS, Barron E, Chui HC (1990) Morphologic association between microglia senile plaque amyloid in Alzheimer's disease. Neurosci Lett 119:32–36

Peterson PK, Hu S, Salak-Johnson J, Molitor TW, Chao CC (1997) Differential production of and migratory response to B chemokines by human microglia and astrocytes. J Infect Dis 175:478–481

Raley MJ, Lennartz MR, Loegering DJ (1999) A phagocytic challenge with IgG-coated erythrocytes depresses macrophage respiratory burst and phagocytic function by different mechanisms. J Leukoc Biol 66:803–806

Rogers J, Lue LF (2001) Microglial chemotaxis, activation, and phagocytosis of amyloid β peptide as linked phenomena in Alzheimer's disease. Interntl J Neurochem (in press)

Rogers J, Luber-Narod J, Styren Sd, Civin WH (1988) Expression of immune system-associated antigens by cells of the human central nervous system: relationship to the pathology of Alzheimer's disease. Neurobiol Aging 9:339–349

Rogers J, Cooper NR, Schultz J, McGeer PL, Webster S, Styren SD, Civin WH, Brachova L, Bradt B, Ward P, Lieberburg I (1992) Complement activation by β-amyloid in Alzheimer's disease. Proc Nat Acad Sci (USA) 89:10016–10020

Rozovsky I, Finch CE, Morgan TE (1998) Age-related activation of microglia and astrocytes: In vitro studies show persistent phenotypes of aging, increased proliferation, and resistance to down-regulation. Neurobiol Aging 19:97–103

Schenk D, Borbour R, Dunn W, et al (1999) Immunization with amyloid-beta attenuates Alzheimer-disease-like pathology in the PDAPP mouse. Nature 400:173–177

Stoltzner SE, Grenfell TJ, Mori C, Wisniewski KE, Wisniewski TM, Selkoe DJ, Lemere C (2000) Temporal accrual of complement proteins in amyloid plaques in Down's syndrome with Alzheimer's disease. Am J Pathol 156:488–499

Sutterwala FS, Noel GJ, Salgame P, Mosser DM (1998) Reversal of proinflammatory responses by ligating the macrophage Fc gamma receptor type I. J Exp Med 188:217–222

Tomozawa Y, Inoue T, Takahashi M, Adachi M, Satoh M (1996) Apoptosis of culturedmicroglia by the deprivation of macrophage colony-stimulating factor. Neurosci Res 25:7–15

Turrin NP, Plata-Salaman CR (2000) Cytokine-cytokine interactions and the brain. Brain Res Bulletin 51:3–9

Ulvestad E, Williams K, Bjerkvig R, Tiekotter K, Antel J, Matre R (1994a) Human microglial cells have phenotypic and functional characteristics in common with both macrophages and dendritic antigen-presenting cells. J Leukoc Biol 56:732–740

Ulvestad E, Williams K, Matre R, Nyland H, Olivier A, Antel J (1994b) Fc receptors for IgG on cultured human microglia mediate cytotoxicity and phagocytosis of antibody-coated targets. J Neuropathol Exp Neuro 53:27–36

Veerhuis R, Janssen I, De Groot CJA, Van Muiswinkel FL, Hack CE, Eikelenboom P (1999) Cytokines associated with amyloid plaques in Alzheimer's disease brain stimulate human glial and neuronal cell cultures to secrete early complement proteins, but not C1-inhibitor. Exp Neurol 160:289–299

Walker DG, Kim SU, McGeer PL (1995) Complement and cytokine gene expression in cultured microglia derived from postmortem human brains. J Neurosci Res 40:478–493

Webster S, Rogers J (1996) Relative efficacies of amyloid β peptide binding proteins in Aβ aggregation. J Neurosci Res 46:58–66

Webster S, Bonnell B, Rogers J (1997a) Charge based binding of complement component C1q to the Alzheimer amyloid β peptide. Am J Pathol 150:1531–1536

Webster S, Bradt B, Rogers J, Cooper NR (1997b) Aggregation state-dependent activation of the classical complement pathway by the amyloid β peptide. J Neurochem 69:388–398

Webster S, Lue L-F, Brachova L, Tenner A, McGeer PL, Walker D, Bradt B, Cooper NR, Rogers J (1997c) Molecular and cellular characterizaiton of the membrane attack complex, C5b-9, in Alzheimer's disease. Neurobiol Aging 18:415–421

Webster SD, Tenner AJ, Poulos TL, Cribbs DH (1999) The mouse C1q A-chain sequence alters beta-amyloid-induced complement activation. Neurobiol Aging 20:297–304

Whittemore SR, Sanon HR, Wood PM (1993) Concurrent isolation and characterization of oligodendrocytes, microglia and astrocytes from adult human spinal cord. Int J Dev Neurosci 11:755–764

Williams K, Bar-Or A, Ulvestad E, Olivier A, Antel J, Yong VW (1992) Biology of adult human microglia in culture: comparisons with peripheral blood monocytes and astrocytes. J Neuropathol Exp Neurol 51:538–549

Wujek JR, Dority MD, Frederickson RC, Brunden KR (1996) Deposits of A beta fibrils are not toxic to cortical and hippocampal neurons in vitro. Neurobiol Aging 17:107–113

Xia MQ, Hyman BT (1999) Chemokines/chemokinereceptors in the central nervous system and Alzheimer's disease. J Neurovirol 5:32–41

Yan SD, Roher A, Schmidt AM, Stern DM (1999) Cellular cofactors for amyloid β-peptide-induced cell stress: moving from cell cultures to in vivo. Am J Pathol 155:1403–1410

Yan SD, Roher A, Chaney M, Zlokovic B, Schmidt AM, Stern D (2000) Cellular cofactors potentiating induction of stress and cytotoxicity by amyloid beta-peptide. Biochim Biophys Acta 1502:145–157

Yan SD, Zhu H, Fu J, Yan SF, Roher, Tourtellotte WW, Ragavashisth T, Chen X, Godman GC, Stern D, Schmidt AM (1997) Amyloid-beta peptide-receptor for advanced glycation endproduct interaction elicits neuronal expres

4 Neuronal SLC (CCL21) Expression: Implications for the Neuron–Microglial Signaling System

K. Biber, A. Rappert, H. Kettenmann, N. Brouwer, S.C.V.M.Copray, H.W.G.M. Boddeke

4.1 Neuron–Microglia Signaling ... 45
4.2 Chemokines and Their Functions in the Immune System ... 46
4.3 Chemokines in Neuroimmunology ... 48
4.4 Expression of CCL21 in the CNS ... 49
4.5 Glial Cells (Microglia and Astrocytes) Express Functional CCL21 Receptors ... 49
4.6 Effects of CCL21 on Cultured Microglia Cells Are Mediated by CXCR3 ... 51
4.7 Function of CCL21 Expression in Neuronal Tissue ... 53
4.8 Conclusions ... 54
References ... 55

4.1 Neuron–Microglia Signaling

It is widely accepted today that glial cells play an important role in many aspects of physiology but also pathology of the CNS. Glia cells represent more than 70% of the cell population of the CNS and consist of three different cell types: astrocytes, oligodendrocytes, and microglia. Particularly, astrocytes and microglia are considered as the neuroimmune cells in the CNS. The microglial cell, originally described by del Rio-Hortega (1932), has a key function in neuroimmunology since microglia are the first cells in brain that are activated in disease and initiate an immune response to nerve injury (Kreuzberg 1996; Raivich et al. 1999). Microglia are bone marrow-derived monocytic cells that invade the brain before establishment of the blood–brain barrier and develop from a typical macrophage-like cell into resting or ramified microglial cells. Ramified microglia are characterized by small, long cell bodies, which are highly branched with tiny, extended processes. Irrespective of the stimulus, activation of microglia proceeds via a stereotypic multi-step process. The first morphological changes are stepwise retractions of their fine processes into a macrophage like, phagocytic cell. Activated microglia are characterized by expression and release of a large variety of immunological mediators such as major histocompatibility complex, complement proteins, prostanoids, nitric oxide, and cytokines, but also neurotrophins and other beneficial molecules (Moore and Thanos 1996; Minghetti and Levi 1998; Frade and Barde 1999; Bruce-Keller 1999; Raivich et al. 1999; Streit et al. 1999). Thus, considerable evidence has accumulated in the last few years that microglia activation not only can be harmful to brain tissue but also may have significant protective functions (Minghetti and Levi 1998; Bruce-Keller 1999; Batchelor et al. 1999; Raivich et al. 1999). Therefore, intensive communication between damaged neurons and microglia, that drives microglial activation and initiates either neuroprotective or neurotoxic responses has been suggested (Kreutzberg 1996; Aschner et al. 1999; Aldskogius and Kozlova 1999; Streit et al. 1999). Little is yet known, however, on the molecular signals participating in neuron–microglial communication. Recently evidence has accumulated that chemokines, a family of small proteins that has been intensively studied in the last decade, are involved in neuron–microglial communication.

4.2 Chemokines and Their Functions in the Immune System

Chemokines constitute a superfamily of small proteins (8–14 kDa) that are instrumental for trafficking of leukocytes in normal immunosurveillance as well as the coordination of infiltration of inflammatory cells under pathological conditions. Currently, approximately 50 different human chemokines have been described, interacting with 18 different chemokine receptors (Murphy et al. 2000; Zlotnik and Yoshie 2000).

Chemokines are classified by their structure based on the number and spacing of conserved cysteine motifs in their sequence. Accordingly, four groups named the C, CC, CXC, and CX3C families have been distinguished. The classification of the chemokine receptors parallels the four subgroups designated for chemokines. These receptor subgroups have been designated XCR, CCR, CXCR, and CX3CR. Chemokines bind to seven transmembrane spanning receptors (Murphy et al. 2000) and activate heterotrimeric G-proteins, which are mostly pertussis toxin sensitive $G_{\alpha i}$-proteins. Thus, the signal transduction of most chemokine receptors often involves inhibition of cyclic adenosine monophosphate (cAMP) and transient increases in intracellular calcium. Furthermore, down-stream activation of mitogen-activated protein kinases (MAPK), phosphoinositide 3-kinase (PI 3-K) and small guanosine triphosphate (GTP)-binding proteins have been described (Sanchez-Madrid and del Polo 1999; Thelen 2001).

A distinction can be made between homeostatic and inducible chemokines. Homeostatic chemokines are mainly involved in physiological traffic-like immune surveillance and homing and predominately expressed in lymphoid organs. Inducible chemokines are induced in various organs during inflammation and orchestrate the infiltration by leukocytes (Mackay 2001). It is now becoming increasingly clear that distinct migration activity of lymphocytes, and presumably also monocytes and granulocytes at various stages of activation, are regulated by complex combinations of homeostatic and inducible chemokines (Moser and Loetscher 2001). In addition to chemotaxis, a number of other biological functions of chemokines such as induction of cell adhesion, phagocytosis, T cell differentiation and activation, apoptosis, angiogenesis, proliferation, and cytokine secretion have been observed (Mackay 2001). The major difference between the chemokine system and other cytokines is the high degree of specificity of chemokine signaling and their ability to fine tune cellular responses within the immune system compared to the rather wide spectrum of actions of most other cytokines (Mackay 2001).

4.3 Chemokines in Neuroimmunology

It has been shown that chemokines and their receptors are expressed in the central nervous system (CNS) during development and pathology (Asensio and Campbell 1999; Glabinski and Ransohoff 1999; Bacon and Harrison 2000; De Groot and Woodroofe 2001). It thus has been suggested that induction of chemokines in the CNS is prominently involved in infiltration of the CNS by blood leukocytes (Ransohoff and Tani 1998; Asensio and Campbell 1999; Menniken et al. 1999; Wu et al. 2000). Recent evidence investigating the induction of experimental autoimmune encephalitis (EAE), which is an animal model for multiple sclerosis (MS), has corroborated this assumption. In recent publications it has been shown that in chemokine/chemokine receptor knockout models the induction of EAE was impaired and was correlated with a pronounced inhibition of infiltration of the CNS by leukocytes (Fife et al. 2000; Izikson et al. 2000; Huang et al. 2001). The chemokines that control the infiltration of the CNS are predominantly produced by astrocytes and microglia.

Furthermore, it has been reported that all cells of the CNS (neurons and all glial cell types) express functional chemokine receptors, which has raised the question on the function of these chemokine receptors (Glabinski and Ransohoff 1999; Hesselgesser and Horuk 1999; Xia and Hyman 1999). Given the wide spectrum of chemokine signaling in the immune system it has been suggested that CNS chemokines, in addition to immune activity, may contribute to an intercellular signaling system involved in pathological processes within the CNS (Hesselgesser and Horuk 1999; Bacon and Harrison 2000).

The possible role of chemokines in neuron–microglia signaling has been proposed since the first chemokine/chemokine receptor (CX3CL1/CX3CR) pair was found in neurons and microglia, respectively (Harrison et al. 1998). Further recent publications showing the expression of other chemokines like IP-10 (CXCL10) or monocyte chemotactic peptide (MCP)-1 (CCL2) in damaged neurons corroborate this assumption (Wang et al. 1998; Che et al. 2001; Fluegel et al. 2001; Schreiber et al. 2001).

4.4 Expression of CCL21 in the CNS

CCL21 is constitutively expressed in secondary lymphoid organs like lymph nodes and Peyer's patches and controls the homing of mature dendritic cells and naïve T cells, all of which express the corresponding chemokine receptor CCR7 (Dieu et al. 1998; Sallusto et al. 1998; Willimann et al. 1998; Yanagihara et al. 1998; Yoshida et al. 1998). CCL21, therefore, has an important function for the immunosurveillance and belongs to the group of "lymphoid" chemokines, meaning that its expression in lymphoid organs is not dependent on the presence of an inflammatory stimulus. CCL21 expression in CNS has not been reported so far.

We have recently shown that ischemia in mice CNS following middle cerebral artery-occlusion (MCAO) induces expression of CCL21 mRNA and protein. CCL21 expression occurred only on the ischemic side of the brain indicating its direct induction by ischemic conditions. Further experiments revealed that the induction of secondary lymphoid-tissue chemokine (SLC) mRNA expression occurs very fast (<6 h) both in vivo and in vitro. Moreover, it was demonstrated that cortical neurons and not glial cells are the cellular source in the CNS of CCL21 expression (see for an overview: Fig. 1 and Biber et al. 2001).

4.5 Glial Cells (Microglia and Astrocytes) Express Functional CCL21 Receptors

The surprising finding of CCL21 expression in ischemic cortical neurons raised questions concerning the function of CCL21 in CNS. Therefore, experiments in order to find possible receptors for CCL21 in brain tissue have been conducted. CCR7, which is the main receptor for CCL21, was neither expressed in healthy nor in ischemic CNS tissue. However, it was shown that an alternative CCL21 receptor, CXCR3 is expressed in glial cells (astrocytes and microglia) (see for an overview: Fig. 1 and Biber et al. 2001; K. Biber et al., submitted). Various experiments using the specific ligands CXCL9 and CXCL10 revealed functional activity of glial CXCR3 (K. Biber et al., submitted).

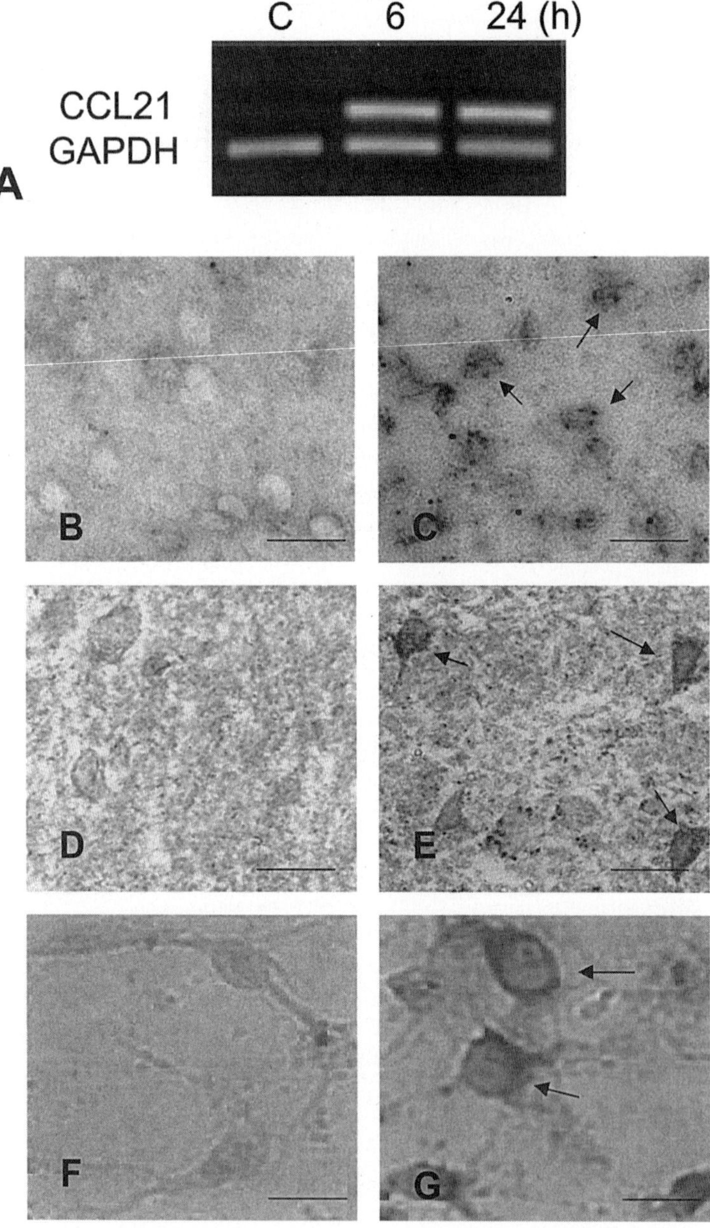
C
6
24 (h)
CCL21
GAPDH
A
B
C
D
E
F
G

4.6 Effects of CCL21 on Cultured Microglia Cells Are Mediated by CXCR3

The first data that CCL21 could act as a chemokine ligand for CXCR3 have been published by Soto and colleagues (Soto et al. 1998). This has been the first and only example indicating that a CCL chemokine can activate a CXC receptor. Currently, the discussion whether CCL21 is a "realistic" chemokine ligand for CXCR3 (compare: Jenh et al. 1999; Lu et al. 1999) continues. Several lines of evidence indicate that murine CCL21 is a chemokine ligand for CXCR3 in murine microglia. (1) CCL21 induces microglial intracellular calcium transients and chemotaxis. (2) The effects of CCL21 on microglial intracellular calcium transients cross-desensitize with CXCL10 (see Fig. 2 for an overview; Biber et al. 2001). Preliminary experiments using microglia cultured from several chemokine receptor knockout mice corroborated this assumption (A. Rappert, K. Biber, H.W.G.M. Boddeke and H. Kettenmann, unpublished findings).

◀

Fig. 1A–G. Expression of CCL21 in ischemic CNS tissue and cultured cortical neurons. **A** RT-PCR analysis of CCL21 mRNA in CNS tissue. CCL21 mRNA expression was not detectable in control CNS tissue but induced by MCAO within 6 h and still prominent 24 h after the occlusion. **B**, **C** In situ hybridization showing the absence of CCL21 mRNA positive cells in the non-ischemic control side of the brain (**B**), but pronounced CCL21 mRNA expression in numerous cells of the ischemic side (**C**) (see *arrows* for examples of CCL21 mRNA positive cells). **D**, **E** CCL21 antibody stainings indicated the expression of CCL21 in cells with neuronal morphology in the ischemic side of the CNS (**E**) and the absence of CCL21 positive cells in the non-ischemic control side (**D**) (see *arrows* for examples of CCL21 positive cells). **F**, **G** In situ hybridization showing the absence of CCL21 mRNA expression in untreated cultured cortical neurons (**F**) and the induction of CCL21 mRNA expression in these cultures by treatment with high concentration of glutamate (**G**) (see *arrows* for examples of CCL21 mRNA positive cells). For details see Biber et al. 2001. *Bar* in B–E, 100 μm; in F and G, 50 μm

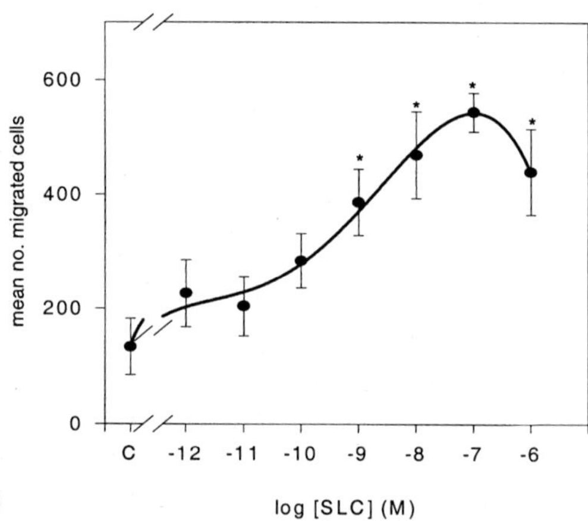

Fig. 2A, B. Effects of CCL21 in cultured microglia. **A** Example of a CCL21 (1 μM)-induced intracellular calcium transient in a cultured microglia cell. *Arrow* indicates the time point of CCL21 application. **B** Effect of CCL21 on chemotaxis of cultured microglia. *, Statistically significant $p<0.01$ (Student's *t*-test). For detail see Biber et al. 2001

4.7 Function of CCL21 Expression in Neuronal Tissue

4.7.1 CCL21 as an Attractor for Brain Infiltrating Leukocytes?

An induction of chemokines has been described in a variety of brain diseases (Glabinski et al. 1995; Gourmala et al. 1998; Balashov et al. 1999; Sorensen et al. 1999; Xia and Hyman 1999). It has been suggested that chemokines mediate infiltration of the brain by leukocytes (Ransohoff and Tani 1998; Asensio et al. 1999; Balashov et al. 1999), which is corroborated by several lines of evidence. Administration of an antibody against the chemokine macrophage inflammatory protein (MIP)-1α (CCL3) inhibited T-cell infiltration in the brain and attenuated the induction of EAE (Karpus et al. 1995). The induction of EAE was also impaired in several chemokine/chemokine receptor knockout models (Fife et al. 2000; Izikson et al. 2000; Huang et al. 2001). Moreover, an early accumulation of monocytes/macrophages in meninges or at later time points in brain parenchyma has been found after injection of the chemokines RANTES (CCL5) (Bell et al. 1996) and C10 (Asensio et al 1999). Injection of CCL21 into mouse cortex did not induce accumulation of monocytes/macrophages in the meninges or brain parenchyma, indicating that SLC is not an attractor for these cells (unpublished findings). These findings are in line with results from other groups showing that SLC does not induce chemotaxis of monocytes/macrophages (Willimann et al. 1998; Nagira et al. 1997, 1998). T cells express CCR7 (Campbell et al. 1998). Neuronally derived SLC would therefore be a potential attractor for T cells in ischemia. However, time point and amount of T-cell infiltration in MCAO does not support this assumption (Schroeter et al. 1994; Jander et al. 1998); moreover, we did not detect CCR7 mRNA expression in mouse brain at any time point after induction of ischemia.

4.7.2 CCL21: A Chemokine Contributing to Neuron–Microglia Signaling

Whereas expression of chemokines in glia cells under neurodegenerative conditions is well established (see for review: Ransohoff and Tani 1998; Glabinski and Ransohoff 1999), relatively few reports show expression of chemokines in neurons. The constitutive neuronal expression of a novel chemokine, named fractalkine (CX3CL1), has been described. Since the corresponding receptor was found to be expressed in microglia, a role of CX3CL1 in neuron–microglia signaling was suggested (Bazan et al. 1997; Pan et al. 1997; Harrison et al. 1998; Nishiyori et al. 1998; Boddeke et al. 1999). Recent evidence suggests that CX3CL1 expression in neurons represents a "homeostatic" chemokine that contributes to the immunosuppressive environment of the CNS and keeps microglia in its ramified, quiescent state (Jung et al. 2000; Zujovic et al. 2000, 2001). Neuronal expression of other chemokines like MCP-1 (CCL2) or CXCL10 is clearly induced by neuronal damage (Wang et al. 1998; Che et al. 2001; Fluegel et al 2001; Schreiber et al. 2001). Like CCL21, these chemokines are induced very rapidly after neuronal damage (within several hours). This is in contrast to chemokine expression in glia cells that normally occurs at later time points (12 h–2 days). It has therefore been suggested that the functions of neuronally expressed chemokines differ from those of glial chemokines, and due to their early expression it is tempting to speculate that chemokines expressed by neurons have fast-acting functions. As mentioned above, one of the first cellular reactions after many types of neuronal injury is the activation of adjacent microglia. Interestingly, it has been published that the first detectable changes in microglia after neuronal damage are changes in their electrophysiology (Boucsein et al. 2000). We therefore asked whether CCL21 stimulation would have an influence on the electrophysiology of microglia. Preliminary experiments substantiated this assumption (A. Rappert, K. Biber, H.W.G.M. Boddeke and H. Kettenmann, unpublished findings).

4.8 Conclusions

CCL21, a chemokine, currently only described in the peripheral immune system, is inducibly expressed in brain, and has been associated with neurodegenerative conditions. Our data show that in brain tissue, neurons are the primary source of CCL21. In contrast to a preference for CCR7 observed in peripheral tissues, in microglia, CCL21 is a primary ligand for CXCR3. Thus, CCL21 expression and receptor interaction in the CNS are different from peripheral immune system.

Microglial cells rapidly respond to any type of CNS injury. Since neuronal chemokines are induced within hours after neuronal damage, chemokines are good candidates for signaling neural injury to microglia. We suggest that CCL21 is involved in direct signaling between neurons and microglia during neuronal injury and presumably participates in early neuroimmune signaling.

References

Aldskogius H, Kozlova EN (1998) Central neuron-glial and glial-glial interactions following axon injury. Prog Neurobiol 55:1–26

Aschner M, Allen JW, Kimelberg HK, LaPachin RM, WJ Streit (1999) Glial cells in neurotoxicity development. Annu Rev Pharmacol Toxicol 39:151–173

Asensio VC, Campbell IL (1999) Chemokines in the CNS: plurifunctional mediators in diverse states. Trends Neuro Sci 22:504–512

Asensio VC, Lassmann S, Pagenstecher A, Steffensen SC, Henriksen SJ, Campbell IL (1999) C10 is a novel chemokine expressed in experimental inflammatory demyelinating disorders that promotes recruitment of macrophages to the central nervous system. Am J Pathol 154:1181–1191

Bacon KB, Harrison JK (2000) Chemokines and their receptors in neurobiology: perspectives in physiology and homeostasis. J Neuroimmunol 104:92–97

Balashov KE, Rottman JB, Weiner HL, WW Hancock (1999) CCR5+ and CXCR3+ T cells are increased in multiple sclerosis and their ligands MIP-1α and IP-10 are expressed in demyelinating brain lesions. Proc Natl Acad Sci 96:6873–6878

Batchelor PE, Liberatore GT, Wong JY, Porritt MJ, Frerichs F, Donnan GA, Howelss DW (1999) Activated macrophages and microglia induce dopaminergic sprouting in the injured striatum and express brain-derived neurotrophic factor and glial cell line-derived neurotrophic factor. J Neurosci 19:1708–1716

Bazan JF, Bacon KB, Hardiman G, Wang W, Soo K, Rossi D, Greaves DR, Zlotnik A, TJ Schall (1997) A new class of membrane-bound chemokine with a CX3C motif. Nature 385:640–644

Bell MD, Taub DD, VH Perry (1996) Overriding the brains intrinsic resistance to leukocyte recruitment with intraparenchymal injections of recombinant chemokines. Neuroscience 74:283–292

Biber K, Sauter A, Brouwer N, Copray JC, Boddeke VM (2001) HWGM Ischemia-induced neuronal expression of the microglia attracting chemokine secondary lymphoid-tissue chemokine (SLC). Glia 34:121–133

Biber K, Dijkstra I, Trebst C, De Groot CJA, Ransohoff RM, Boddeke HWGM (submitted) Functional expression of CXCR3 in cultured mouse and human astrocytes. Microglia

Boddeke HWGM, Meigel I, Frentzel S, Biber K, Ren L, Gebicke-Härter PJ (1999) Functional expression of the fractalkine (CX_3C) receptor and its regulation by lipopolysaccharide in rat microglia. Eur J Pharmacol 374:309–313

Boucsein C, Kettenmann H, Nolte C (2000) Electrophysiological properties of microglial cells in normal and pathologic rat brain slices. Eur J Neurosci 12: 2049–2058

Bruce-Keller AJ (1999) Microglial and neuronal interactions in synaptic damage recovery. J Neurosci Res 58:191–201

Campbell JJ, Bowman EP, Murphy K, Youngman KR, Siani MA, Thompson DA, Wu L, Zlotnik A, EC Butcher (1998) 6-C-kine (CCL21), a lymphocyte adhesion-triggering chemokine expressed by high endothelium, is an agonist for the MIP-3β receptor CCR7. J Cell Biol 141:1053–1059

Che X, Ye W, Panga L, Wu DC, Yang GY (2001) Monocyte chemoattractant protein-1 expressed in neurons and astrocytes during focal ischemia in mice. Brain Res 902:171–177

De Groot CJ, Woodroofe MN (2001) The role of chemokines and chemokine receptors in CNS inflammation. Prog Brain Res 132: 533–544

del Rio-Hortega P (1932) Microglia. In: Penfield W (ed) Cytology and cellular pathology of the nervous system. Paul P. Hocker Inc., New York, pp 481–534

Dieu MC, Vanbervliet B, Vicari A, Bridon JM, Oldham E, Ait-Yahia S, Briere F, Zlotnik A, Lebecque S, Caux C (1998) Selective recruitment of immature and mature dendritic cells by distinct chemokines expressed in different anatomic sites. J Exp Med 188:373–386

Fife BT, Huffnagle GB, Kuziel WA, Karpus WJ (2000) CC chemokine receptor 2 is critical for induction of experimental autoimmune encephalomyelitis. J Exp Med 192:899–905

Fluegel A, Hager G, Horvat A, Spitzer C, Singer GMA, Graeber MB, Kreutzberg GW, Schwaiger F-W (2001) Neuronal MCP-1 expression in response to remote nerve injury. J Cereb Blood Flow Metab 21:69–76

Frade JM, Barde YA (1999) Microglia-derived nerve growth factor causes cell death in the developing retina Neuron 20:35–41

Glabinski AR, Ransohoff RM (1999) Chemokines and chemokine receptors in CNS pathology. J Neurovirol 5:3–12

Glabinski AR, Tani M, Tuohy VK, Tuthill R, Ransohoff RM (1995) Central nervous system chemokine gene expression follows leukocyte entry in acute murine experimental autoimmune encephalomyelitis. Brain Behav Immun 9:315–330

Gourmala NG, Buttini M, Limonta S, Sauter A, Boddeke HW (1998) Differential and time-dependent expression of monocyte chemoatractant protein-1 mRNA by astrocytes and macrophages in rat brain: effects of ischemia and peripheral lipopolysaccharide administration. J Neuroimmunol 74:35–44

Harrison JK, Jiang Y, Chen S, Xia Y, Maciejewski D, McNamara RK, Streit WJ, Salafranca MN, Adhikari S, Thompson D A, Botti P, Bacon KB, L Feng (1998) Role for neuronally derived fractalkine in mediating interactions between neurons and CX3CR1-expressing microglia. Proc Natl Acad Sci 95:10896–10901

Hesselgesser J, Horuk R (1999) Chemokine and chemokine receptor expression in the central nervous system. J Neurovirol 5:13–26

Huang D, Wang J, Kivisakk P, Rollins BJ, Ransohoff RM (2001) Absence of monocyte chemoattractant protein 1 in mice leads to decreased local macrophage recruitment and antigen-specific t helper cell type 1 immune response in experimental autoimmune encephalomyelitis. J Exp Med 93:713–726

Izikson L, Klein RS, Charo IF, Weiner HL, Luster AD (2000) Resistance to experimental autoimmune encephalomyelitis in mice lacking the CC chemokine receptor (CCR)2. J Exp Med 192:1075–1080

Jander S, Schroeter M, D'Urso D, Gillen C, Witte OW, Stoll G (1998) Focal cerebral ischemia of the rat brain elicits an unusual inflammatory response: early appearance of CD8+ macrophages/microglia. Eur J Neurosci 10: 680–688

Jenh CH, Cox MA, Kaminski H, Zhang M, Byrnes H, Fine J, Lundell D, Chou CC, Narula SK, Zavodny P (1999) Species specificity of the CC chemokine 6Ckine signalling through the CXC chemokine receptor CXCR3: Human 6Ckine is not a ligand for the human or mouse CXCR3 receptors. J Immunol 162:765–3769

Jung S, Aliberti J, Graemmel P, Sunshine MJ, Kreutzberg GW, Sher A, Littman DR (2000) Analysis of fractalkine receptor CX_3CR1 function by targeted deletion and green fluorescent protein reporter gene insertion. Mol Cell Biol 20: 1–13

Karpus WJ, Lukacs NW, McRae BL, Strieter RM, Kunkel SL, Miller SD (1995) An important role for the chemokine macrophage inflammatory protein-1 alpha in the pathogenesis of the T cell-mediated autoimmune disease, experimental autoimmune encephalomyelitis. J Immunol 155:5003–5010

Kreutzberg GW (1996) Microglia: a sensor for pathological events in the CNS. Trends Neurosci 19:312–318

Lu B, Humbles A, Bota D, Gerard C, Moser B, Dulce S, Luster AD, Gerard N (1999) Structure and function of the murine chemokine receptor CXCR3. Eur J Immunol 29:3804–3812

Mackay CR (2001) Chemokines: immunology's high impact factors. Nat Immunol 2:95–101

Mennicken F, Maki R, de Souza EB, Quirion R (1999) Chemokines and chemokine receptors in the CNS: a possible role in neuroinflammation and patterning. Trends Pharmacol Sci 20:73–78

Minghetti L, Levi G (1998) Microglia as effector cells in brain damage and repair: Focus on prostanoids and nitric oxide. Prog Neurobiol 54:99–125

Moore S, Thanos S (1996) The concept of microglia in relation to central nervous system disease and regeneration. Prog Neurobiol 35:21–30

Moser B, Loetscher P (2001) Lymphocyte traffic control by chemokines. Nat Immunol 2:123–128

Murphy PM, Baggiolini M, Charo IF, Hebert CA, Horuk R, Matsushima K, Miller LH, Oppenheim JJ, Power CA (2000) International union of pharmacology. XXII. Nomenclature for chemokine receptors. Pharmacol Rev 52:145–76

Nagira M, Imai T, Hieshima K, Kusuda J, Rindanpaa M, Takagi S, Nishimura M, Kakizaki M, Nomiyama H, Yoshie O (1997) Molecular cloning of a novel human CC chemokine Secondary Lymphoid-tissue Chemokine that is a potent chemoattractant for lymphocytes and mapped to chromosome 9p13. J Biol Chem 272:19518–19524

Nagira M, Imai T, Yoshida R, Takagi S, Iwasaki M, Baba M, Tabira Y, Akagi J, Nomiyama H, Yoshie O (1998) A lymphocyte-specific CC chemokine, secondary lymphoid tissue chemokine (CCL21), is a highly efficient chemoattractant for B cells and activated T cells. Eur J Immunol 28:1516–1523

Nishiyori A, Minami M, Ohtani Y, Takami S, Yamamoto J, Kawaguchi N, Kume T, Akaike A, Satoh M (1998) Localization of fractalkine and CX3CR1 mRNAs in rat brain: does fractalkine play a role in signalling from neurons to microglia? FEBS Lett 429:167–172

Pan Y, Lloyd C, Zhou H, Dolich S, Deeds J, Gonzalo J, Vath J, Gosselin M, Ma J, Dussault B, Woolf E, Alperin G, Culpepper J, Gutierrez-Ramos J, Gearing D (1997) Neurotactin a membrane-anchored chemokine upregulated in brain inflammation. Nature 387:611–617

Raivich G, Bohatschek M, Kloss CUA, Werner A, Jones LL, Kreutzberg GW (1999) Neuroglial activation repertoire in the injured brain: Graded response molecular mechanisms and cues to physiological function. Brain Res Rev 30:77–105

Ransohoff RM, Tani M (1998) Do chemokines mediate leukocyte recruitment in post-traumatic CNS inflammation? Trends Neurosci 21:154–159

Sallusto F, Schaerli P, Loetscher P, Schaniel C, Lenig D, Mackay CR, Qin S, Lanzavecchia A (1998) Rapid and coordinated switch in chemokine receptor expression during dendritic cell maturation. Eur J Immunol 28:2760–2769

Sanchez-Madrid F, del Pozo MA (1999) Leukocyte polarization in cell migration and immune interactions. EMBO J 18:501–511

Schreiber RC, Krivacic K, Kirby B, Vaccariello SA, Wei T, Ransohoff RM, Zigmond RE (2001) Monocyte chemoattractant protein (MCP)-1 is rapidly expressed by sympathetic ganglion neurons following axonal injury. Mol Neurosci 12:601–605

Schroeter M, Jander S, Witte OW, Stoll G (1994) Local immune responses in the rat cerebral cortex after middle cerebral artery occlusion. J Neuroimmunol 55:195–203

Sorensen TL, Tani M, Jensen J, Pierce V, Lucchinetti C, Folcik VA, Qin S, Sellebjerg F, Strieter RM, Frederiksen JL, Ransohoff RM (1999) Expression of specific chemokines and chemokine receptors in the central nervous system of multiple sclerosis patients. J Clin Invest 103:807–815

Soto H, Wang W, Strieter RM, Copeland NG, Gilbert DJ, Jenkins NA, Hedrick J, Zlotnik A (1998) The CC chemokine 6Ckine binds the CXC chemokine receptor CXCR3. Proc Natl Acad Sci 95:8205–8210

Streit WJ, Walter SA, Pennell NA (1999) Reactive microgliosis. Prog Neurobiol 57:563–581

Thelen M (2001) Dancing to the tune of chemokines. Nat Immunol 2:129–34

Wang X, Ellison JA, Siren A-L, Lysko PG, Yue T-L, Barone FC, Shatzman A, Feuerstein GZ (1998) Prolonged expression of interferon-inducible protein-10 in ischemic cortex after permanent occlusion of the middle cerebral artery in rat. J Neurochem 71:1194–1204

Willimann K, Legler DF, Loetscher M, Roos RS, Delgado MB, Clark-Lewis I, Baggiolini M, Moser B (1998) The chemokine SLC is expressed in T cell areas of lymph nodes and mucosal lymphoid tissues and attracts activated T cells via CCR7. Eur J Immunol 28:2025–2034

Wu DT, Woodman SE, Weiss JM, McManus CM, D'Aversa TG, Hesselgesser J, Major EO, Nath A, Berman JW (2000) Mechanisms of leukocyte trafficking into the CNS. J Neurovirol [Suppl 1]:82–85

Xia M, Hyman BT (1999) Chemokines/chemokine receptors in the central nervous system and Alzheimer's disease. J Neurovirol 5:32–41

Yanagihara S, Komura E, Nagafune J, Watarai H, Yamaguchi Y (1998) EBI1/CCR7 is a new member of dendritic cell chemokine receptor that is up-regulated upon maturation. J Immunol 161:3096–3102

Yoshida R, Nagira M, Kitaura M, Imagawa N, Imai T, Yoshie O (1998) Secondary Lymphoid-tissue chemokine is a functional ligand for the CC chemokine receptor CCR7. J Biol Chem 273:7118–7122

Zlotnik A, Yoshie O (2000) Chemokines: a new classification system and their role in immunity. Immunity 2:121–127

Zujovic V, Benavides J, Vige X, Carter C, Taupin V (2000) Fractalkine modulates TNF-alpha secretion and neurotoxicity induced by microglial activation. Glia 29:305–315

Zujovic V, Schussler N, Jourdain D, Duverger D, Taupin V (2001) In vivo neutralization of endogenous brain fractalkine increases hippocampal TNFα and 8-isoprostane production induced by intracerebroventricular injection of LPS. J Neuroimmunol 115:135–143

5 Cytokine-Mediated Inflammation and Other Actions in the Central Nervous System

I.L. Campbell

5.1 Introduction 61
5.2 Distinct Neurological Phenotypes Induced by Astrocyte-Targeted Cytokine Production 62
5.3 Molecular Actions of Cytokines in the GFAP-Cytokine Transgenic Mice 67
5.4 Molecular Circuitry for Cytokine Signaling in the CNS 73
5.5 Concluding Remarks 77
References 78

5.1 Introduction

Cytokines are potent autocrine or paracrine regulatory factors in the host response to injury or immunological challenge that regulate a diverse array of functions necessary for tissue and host protection and recovery. In particular, these molecules are pivotal in the regulation of inflammatory processes. In normal physiologic states, the level of various cytokines in the CNS is tightly controlled and very low to undetectable. However, increased CNS expression of various cytokines is found in a variety of neurological disorders such as multiple sclerosis (MS) (Arnason 1995), human immunodeficiency virus (HIV)-1 dementia (HAD) (Lipton 1997), Alzheimer's disease (Rogers et al. 1996), spongiform encephalopathies (Campbell et al. 1994; Williams et al. 1994), viral

(Griffin 1997), bacterial (Leist et al. 1988; Waage et al. 1989) or parasite infection (Suzuki 1999), stroke (Feuerstein et al. 1997) and traumatic injury (Ghirnikar et al. 1998). Importantly, in these different pathologic states, cells that are intrinsic to the CNS, particularly the microglia and astrocytes, have been identified as major cytokine producers. Since cytokines are very pleiotropic and potent regulators of cellular functions involved in inflammation, it has been speculated widely that cytokines might play fundamental roles in the pathogenesis of neuroinflammatory disease. What these roles are and the nature and mechanisms by which cytokines mediate their actions in the CNS, and on what cells, are far from clear.

To properly address these issues, it should first be considered that the CNS is a highly complex, dynamic, and interactive tissue. Therefore, and ideally, the actions of cytokines are most appropriately examined in the living, unperturbed CNS. In this regard, one approach that has been employed extensively by us utilizes a transgenic strategy to target the expression of individual cytokine genes to astrocytes in the CNS of mice (Campbell 1998a,b). This functional genomic approach offers many advantages. First, it permits the prolonged, reproducible delivery of a pure gene product to specific cells in the living CNS where the anatomic and physiologic interactions are preserved. Second, lines of transgenic mice can be generated providing an unrestricted source of animals identical for the introduced genetic alteration. Third, these mutant mice facilitate integrative studies to link molecular, cellular, and higher order functional neurological (e.g., cognitive) outcomes. In this article, I address some general concepts that have emerged and will emphasize the use of CNS-cytokine transgenic models to gain a better understanding of basic mechanisms that underlie cytokine-mediated inflammation in the CNS.

5.2 Distinct Neurological Phenotypes Induced by Astrocyte-Targeted Cytokine Production

For the generation of transgenic mice with astrocyte-targeted expression of cytokine genes we employed fusion gene constructs (Fig. 1) in which the coding region of the cytokine DNA was placed under the transcriptional control of the glial fibrillary acidic protein (GFAP)-promoter.

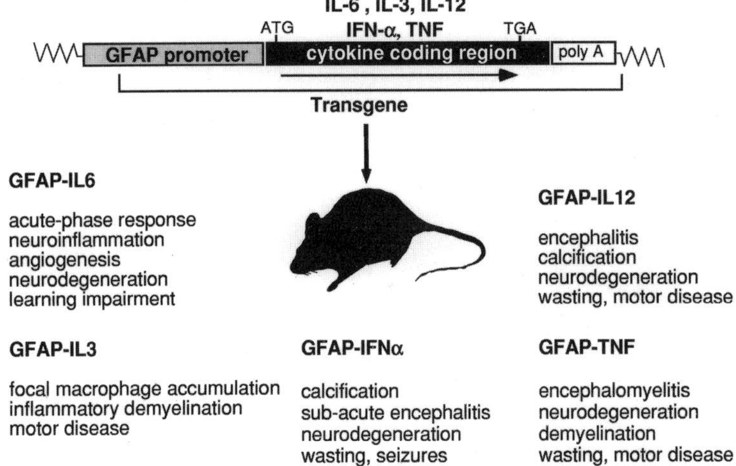

Fig. 1. Overall strategy for the assessment of cytokine-CNS interactions using transgenic modeling. GFAP-cytokine fusion gene constructs were used to develop transgenic mice with astrocyte-targeted cytokine expression. GFAP-cytokine transgenic mice develop a spectrum of structural and functional neurological alterations that are somewhat specific for each cytokine

Table 1. Characteristics of cytokines used in the development of GFAP-transgenic mice

Cytokine	Type	CNS expression	Disease association
IL-6	Neuropoietic/ proinflammatory	Astrocyte, microglia, neuron	Alzheimer's disease, NeuroAIDS, stroke, infection, trauma
IL-3	Hematopoietic/ proinflammatory	?	EAE
IL-12	Proinflammatory	Astrocyte, microglia	Multiple sclerosis, infection
IFN-α	Antiviral/ proinflammatory	Astrocyte, microglia	Multiple sclerosis, NeuroAIDS, infection
TNF	Proinflammatory/ counter-inflammatory	Astrocyte, microglia, neuron	Multiple sclerosis, stroke, bacterial infection, trauma

This strategy has allowed us to successfully direct the production of a number of cytokines to the murine CNS, including interleukin (IL)-3 (Chiang et al. 1996), IL-6 (Campbell et al. 1993), IL-12 (Pagenstecher et al. 2000), interferon (IFN)-α (Akwa et al. 1998), and tumor necrosis factor (TNF) (Stalder et al. 1998). With the possible exception of IL-3, all of these cytokines are known to be produced by glial cells and are implicated in the pathogenesis of different neuroinflammatory disorders (see Table 1). Characteristically, for each of these cytokines, GFAP-encoded expression at low levels in the transgenic mice allowed for the further development of stable breeding lines. Importantly, CNS production of each of these cytokines is documented at both the RNA and protein levels and in general approximates that found in the CNS in experimentally-induced neuroinflammatory states, such as in experimental autoimmune encephalomyelitis (EAE) or during infection. Thus, the production of the cytokines in these transgenic mice, while chronic, is at pathophysiologically relevant levels. Consequently, and with the exception of GFAP-IL-6 mice, overt CNS alterations with neuroinflammation is only observed in older GFAP-cytokine transgenic mice. A brief discussion of some of the more prominent features follows.

The astrocyte-targeted production of three cytokines, viz., IL-3, IL-12 and TNF, induces robust immunoinflammatory responses and degenerative disease in the brain. Typically, these responses occur in animals 5 months of age or older and are associated with progressive motor impairment and wasting, leading to premature death. The brains of affected mice have large numbers of infiltrating leukocytes that accumulate in perivascular and parenchymal regions of the CNS, overlapping with the principal sites of transgene-encoded cytokine production. The cellular composition of the leukocytic infiltrates and the associated degenerative pathology varies for the different cytokines.

In the case of IL-3, which is known to stimulate the activation and proliferation of macrophages (Frendl and Beller 1990) as well as microglia (Lee et al. 1993), astrocyte-targeted production of this cytokine induces the accumulation of activated and proliferating macrophage/microglia in hypercellular lesions in the white matter with only small numbers of $CD4^+$ T cells being present (Chiang et al. 1996). Most of the accumulating macrophage/microglia are $CD45^{high}$ suggesting an origin from peripheral monocytes rather than from resident microglia (Carson et al. 1999). These lesions are associated with primary demyelination, oligodendrocyte destruction, and subsequent axonal injury. This pathologic process appears to be mediated by the macrophage/microglia directly since direct stripping by these cells of axonal myelin sheath has been observed. The $CD4^+$ T cells in these lesions are not necessary for the development of the macrophage/microglial-mediated inflammatory demyelinating disease, since GFAP-IL-3 X severe combined immunodeficiency (SCID) mice develop the same inflammatory demyelinating phenotype (I.L. Campbell, unpublished finding). The molecular and cellular pathologic manifestations exhibited by the GFAP-IL-3 mice show overlap with the active plaque lesions of MS highlighting the potential role of macrophage/microglia in the pathogenesis of this demyelinating disorder.

In contrast to IL-3, inflammatory lesions in the GFAP-IL-12 (Pagenstecher et al. 2000) and -TNF (Stalder et al. 1998) mice contain high numbers of $CD4^+$ and $CD8^+$ T-cells. In addition, NK-cells or B cells and neutrophils are also numerous in GFAP-IL-12 or -TNF mice, respectively. Analysis of the phenotypic properties of the infiltrating T cells by flow cytometry showed there were significant differences between these models. Thus, whereas in the GFAP-IL-12 mice CNS infiltrating T cells are highly activated, those from the GFAP-TNF mice are not and have the property of resting memory T cells. Functional differences in the T-cell compartment are also apparent between these two models with the presence of IFN-γ gene expression in the CNS of GFAP-IL-12, but not -TNF mice. These findings indicate that while astrocyte production of both IL-12 and TNF can promote T-cell recruitment and accumulation in the CNS, only IL-12 is capable of activating these immune cells. Despite this, generalized degenerative changes are seen in both the GFAP-IL-12 and -TNF brain with frank neurodegeneration and demyelination occurring in close association with the immunoinflammatory lesions. Calcification is an additional prominent pathological feature found in the brain of GFAP-IL-12 but not -TNF mice. The role of the T and B cells in mediating the pathological changes in the GFAP-TNF mice has been assessed in the SCID background. Surprisingly, GFAP-TNF/SCID mice that lack T and B lymphocytes develop severe motor disease with an earlier onset and more extensive inflammatory lesions, indicating that the lymphocytes are not necessary, and may even be protective for TNF-induced neuroinflammatory disease (Stalder et al. 1998). The CNS lesions in the GFAP-TNF/SCID mice contain predominantly activated macrophage/microglia, highlighting a dramatic direct response by these inflammatory cells to chronic TNF production. The findings again underscore the central role of macrophage/microglia as mediators of destructive inflammatory disease in the CNS. Similar studies employing SCID mice are currently underway to determine the role of T cells in the neuroimmune disease phenotype of GFAP-IL-12 mice.

Infiltration of the brain with leukocytes is far less dramatic in transgenic mice with astrocyte-targeted production of the cytokines IFN-α (Akwa et al. 1998; Campbell et al. 1999) and IL-6 (Campbell et al. 1993; Chiang et al. 1994; Heyser et al. 1997). Nevertheless, a progressive inflammatory phenotype with concomitant neurodegenerative disease develops in the CNS of these animals. In general, GFAP-IFN-α and -IL-6 transgenic mice become progressively moribund, are prone to seizures, and die prematurely. In spite of the similar physical phenotypes, the pathological changes are quite disparate between these two models. In GFAP-IFN-α255D mice, infiltrating cells consist predominantly of low numbers of $CD4^+$ T cells around vessels and in the meninges, while in GFAP-IL-6 mice B cells predominate. A very conspicuous pathological feature in GFAP-IFN-α mice not found in the brain of GFAP-IL-6 mice is progressive calcification affecting basal ganglia and cerebella granule neuron layers. Lack of proximity to the T-cell infiltrates suggests this calcification is independent of any immune process and likely reflects a direct action of the IFN-α. This contrasts with the GFAP-IL-12 mice, where calcification only occurs in concert with the immunoinflammatory disease. Interestingly, as noted above, the immunoinflammatory process in the GFAP-IL-12 mice is accompanied by the expression of IFN-γ in the brain making it likely that it is this cytokine, rather than IL-12, that is responsible for the calcification. It seems probable that IFN-γ and IFN-α may mediate calcification in the CNS via overlapping signaling pathways. In addition to calcification, another notable difference between the GFAP-IFN-α and -IL-6 mice involves alterations to the cerebrovasculature with marked degeneration and dilation of vessels seen in the former, while progressive angiogenesis occurs in the latter. While the resultant neurodegeneration in the GFAP-IFN-α mice affects predominantly cholinergic neurons, in the GFAP-IL-6 mice it does not, but rather, in this model it involves loss of parvalbumin- and calbindin-positive interneurons. Interestingly, a progressive deterioration in learning and memory in the GFAP-IL-6 mice correlates significantly with the loss of the calbindin interneurons and the degree of microglial activation (Heyser et al. 1997).

The foregoing discussion makes a number of points clear. First, this transgenic approach allows the assessment of the impact of individual cytokines on the living CNS and, as such, the findings document that these molecules alone can directly mediate neuroinflammatory disease. Second, in reflecting upon the specific outcomes for each cytokine, the structural and functional inflammatory phenotypes of the different GFAP-cytokine transgenic mice are somewhat distinct. A summary of the predominant features of these different GFAP-cytokine transgenic models is given in Fig. 1. Third, the neuropathological phenotypes displayed by the GFAP-cytokine transgenic mice often mirror those reported in human neurological disorders. One striking example is illustrated by the encephalopathy in GFAP-IFN-α mice, which shows great similarity to the human familial encephalopathies Aicardi-Goutières syndrome (Aicardi and Goutieres 1984; Tolmie et al. 1995) and Cree encephalitis (Black et al. 1988). Significantly, both human disorders are associated with increased intrathecal production of IFN-α (Black et al. 1988; Lebon et al. 1988). Thus, the murine transgenic model offers direct evidence that inappropriate or chronic IFN-α production in the CNS may be a primary causative factor for Aicardi-Goutières syndrome and Cree encephalitis. Both Aicardi-Goutières syndrome and Cree encephalitis may, therefore, represent the first known examples of primary cytokine encephalopathies.

5.3 Molecular Actions of Cytokines in the GFAP-Cytokine Transgenic Mice

As we have seen from the above discussion, CNS-targeted expression of different cytokines produces distinct inflammatory and functional responses. This raises the question, What are the molecular mechanisms that cause the individual cytokine-induced actions in the brain? Despite evidence for intrinsic resistance of the CNS to the development of immunoinflammation, recruitment and accumulation of leukocytes in the CNS is a prominent response to many of the cytokine genes we have expressed in the astrocytes. Therefore, we have examined the underlying molecular basis for this response.

Leukocyte trafficking is known to be a multifactorial process involving a number of key molecular components including the cellular adhesion molecules (CAMs), chemokines and matrix metalloproteinases (MMPs). The CAMs are critical for the initial binding of leukocytes to the vascular endothelium and facilitate their subsequent extravasation at sites of inflammation (Butcher 1990). Important CAMs include intercellular adhesion molecule-1 (ICAM-1), vascular cellular adhesion molecule-1 (VCAM-1) and the mucosal addressin cellular adhesion molecule (MAdCAM). ICAM-1 is found at low levels on vascular endothelium in the CNS (Steffen et al. 1996). Marked upregulation in the expression of ICAM-1 occurs on cerebrovascular endothelium, choroid plexus, and on infiltrating leukocytes in the GFAP-TNF, -IFN-α and -IL-12 transgenic mice. More modest increases in ICAM-1 expression are also found in the GFAP-IL-6 mice while little or no detectable change in the expression of this CAM is seen in the GFAP-IL-3 mice. While the levels of VCAM-1 are very low to absent in brain from wild-type mice, markedly increased expression of this CAM is found on large and small vessels in the GFAP-TNF and -IL-12 mice and on small vessels in the GFAP-IFN-α transgenic mice. No detectable alteration in VCAM-1 expression is apparent in the CNS of the GFAP-IL-6 or -IL-3 mice. Expression of MAdCAM is normally quite restricted (Butcher 1990) and is absent from the CNS of wild-type mice. However, only in GFAP-TNF mice is the induction of this CAM observed. Thus, there are wide differences in the expression of the individual CAMs by the cerebrovascular endothelium in the different GFAP-cytokine transgenic mice ranging from upregulation or induction of all three studied molecules in the GFAP-TNF mice to little or no detectable alteration in the GFAP-IL-3 mice. The exact significance of these differences is unknown, but conceivably, they might serve to regulate the phenotypic makeup of infiltrating leukocytes in the brain of the different GFAP-cytokine transgenic mice.

Chemokines are chemoattractant cytokines involved in the tissue recruitment of leukocytes in inflammatory states (Schall and Bacon 1994; Haelens et al. 1996; Taub 1996). The chemokines are composed of four distinct subfamilies grouped according to their NH_2-terminal cysteine structural motif being CXC (the α-subfamily), CC (the β-subfamily), C (the γ family) or CX3C (the δ subfamily) where X is the intervening amino acid residue. Chemokines are now recognized as having central roles in regulating leukocyte migration to the CNS in inflammatory disease states (Ransohoff et al. 1996; Asensio and Campbell 1999). In the brain from wild-type mice, with the exception of the CX3C chemokine fractalkine and the CXC chemokine SDF-1, there is little to no detectable expression of most members of the various chemokine families. In contrast, cerebral chemokine gene expression in the GFAP-IL-3, -IL-6, -IFNα or -TNF transgenic mice show considerable differences in the levels, patterns, and cellular localization (Stalder et al. 1998; Asensio et al. 1999). In the GFAP-IL-3 mice, high expression of the CC chemokine gene C10, is found restricted to macrophage/microglia, while in the GFAP-IFNα mice, the CC chemokine gene, IP-10, is expressed at high levels by astrocytes. In contrast, in GFAP-IL-6 mice only very low levels of IP-10 and RANTES gene expression are detectable. By contrast, in the GFAP-TNF mice there is overlapping and high CNS expression of a number chemokine genes belonging to both the CC (IP-10 and MIP-2) and CC (MCP-1 and RANTES) chemokine families. These differential patterns of chemokine gene expression reflect the overall complexity of the leukocyte populations found in the brain of the GFAP-cytokine. Thus, in GFAP-IL-3 and IFN-α mice, CNS infiltrates are largely composed of a single class of leukocyte and the repertoire of chemokine genes expressed is limited. However, in GFAP-TNF mice which have dense infiltrates consisting of mononuclear cells and polymorphonuclear cells, the repertoire of expressed chemokine genes is broad. The expression of these specific chemokines in these different inflammatory states likely directly controls the type of leukocyte that is recruited to the CNS. In support of this, C10 was shown to be effective at promoting macrophage recruitment to the mouse brain following intracerebroventricular injection (Asensio et al. 1999). Induction of macrophage/microglial C10 expression by IL-3 in the brain of the GFAP-IL-3 mice appears to be a key response that in turn leads to increased macrophage recruitment from the periphery.

To migrate from the vascular compartment through the parenchyma of tissues, leukocytes utilize MMPs which have the capacity to digest and remodel the extracellular matrix (ECM) (Goetzl et al. 1996). The activity of the MMPs is counterregulated by the tissue inhibitor of the matrix metalloproteinases (TIMPs). MMPs and TIMPs are implicated in the pathogenesis of inflammatory disorders of the CNS (Yong et al. 1998; Campbell and Pagenstecher 1999). We examined the temporal and spatial expression patterns of a number of MMP genes as well as the tissue inhibitor of the TIMPs genes in the CNS of the GFAP-IL-3, -IL-6 or -TNF transgenic mice (Pagenstecher et al. 1998). In wild-type mice, the MMPs MT1-MMP (MMP-14), stromelysin 3 (MMP-11) and gelatinase B (MMP-9) are expressed at low levels, while high expression of TIMP-2 and TIMP-3 is found predominantly in neurons and in the choroid plexus, respectively. In the transgenic mice, while there is little alteration in MMP gene expression in the GFAP-IL-6 mice, marked upregulation in macrophage metalloelastase (MMP-12) occurs in CNS infiltrating macrophage/microglia in the GFAP-IL-3 and -TNF mice. In GFAP-TNF mice, there is also increased CNS expression of gelatinase A (MMP-2) and stromelysin 1. Thus, similar to the chemokines, the MMP expression patterns reflect the overall cellular complexity of the immunoinflammatory lesions. In particular, these studies identify a potentially important role for the macrophage metalloelastase (MMP-12) in the CNS infiltration/migration by macrophage/microglia and perhaps the subsequent tissue destruction mediated by these cells.

Interestingly, in association with the increased MMP gene expression, in all the transgenic models there is a significant induction of TIMP-1 gene expression in the CNS (Pagenstecher et al. 1998). There is a clear dichotomy in the cellular localization of these molecules with MMP expression being restricted to leukocytes and possibly microglia within inflammatory lesions, while TIMP-1 is found almost exclusively in activated astrocytes circumscribing the lesions. Thus, there is specific spatial and temporal regulation in the expression of individual MMP and TIMP genes in the CNS in normal and the inflammatory states produced by the astrocyte-targeted production of these cytokines. This suggests the interactions between these gene products may determine both the size and resolution of the destructive inflammatory focus.

The foregoing discussion makes the important point that the diverse immunoinflammatory phenotypes induced by the astrocyte-targeted expression of these different cytokines arise from the differential regulation of key molecular targets that regulate leukocyte trafficking and activity. The same can certainly be said for other phenotypes that we have identified in the GFAP-cytokine transgenic mice. This is vividly

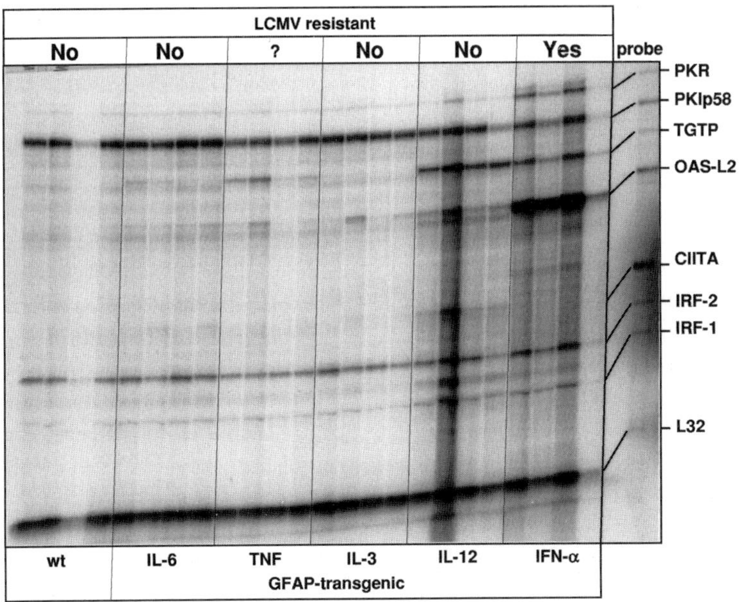

Fig. 2. Distinct functional and molecular responses to the astrocyte-targeted production of different cytokines in GFAP-cytokine transgenic mice. The expression of a number of IFN-regulated genes in the brain was analyzed by multiprobe RNase protection assay as described previously (Asensio et al., 2001). Significant variation in the expression of these different genes can be seen with marked upregulation of the RNA-dependent protein kinase (*PKR*), T-cell GTPase (*TGTP*) and 2′5′-oligoadenylate synthetase L2 gene (*OAS-L2*) in GFAP-IFN-α mice. In comparison, in GFAP-IL-12 mice, TGTP gene expression is similarly increased, however, expression of the PKR and OAS-L2 genes is considerably less. Only minor changes in the expression of all these genes can be seen in GFAP-IL-6, -IL-3 and -TNF mice. In parallel with these molecular changes, GFAP-IFNα mice are very resistant to lethal LCMV infection. In contrast, GFAP-IL-6, -IL-3 and -IL-12 mice show little or no resistance to LCMV infection

illustrated in the case of infection with a neurotropic virus such as lymphocytic choriomeningitis virus (LCMV). Intracranial infection of wild-type, GFAP-IL-3, -IL-6 and -IL-12 mice with LCMV results in the development of severe LCM and death of all infected animals. In contrast, similar infection in GFAP-IFN-α mice produces only a mild LCM and the majority of infected animals survive (Akwa et al. 1998). Consistent with the well-known antiviral actions of IFN-α, the resistance of GFAP-IFN-α mice to fatal LCMV infection is associated with a profound reduction in the levels of LCMV present in the brain following infection. We have recently explored the molecular basis for the differences in the LCMV response in the different groups of mice, focusing on a number of IFN-regulated genes, many of which are known to mediate the antiviral actions of the IFNs. As shown in Fig. 2, in the brain of GFAP-IFN-α mice there is significantly increased expression of a variety of genes not found in wild-type or the other GFAP-cytokine transgenic mice. In particular, the RNA-dependent protein kinase (PKR), 2′5′-oligoadenylate synthetase (OAS-L2) and T-cell guanosine triphosphate (GTP)ase (TGTP) are known to have antiviral actions that alone or together likely contribute to the LCMV-resistant phenotype of the GFAP-IFN-α mice. It should be noted that despite the presence of IFN-γ in the brain, the GFAP-IL-12 mice do not exhibit an LCMV-protected phenotype and have only very modest alterations in PKR, OAS-L2, and TGTP cerebral gene expression. These observations highlight the remarkable specificity in the actions of the type I (e.g., IFN-α) versus type II (i.e., IFN-γ) IFNs in the CNS.

When comparing the GFAP-cytokine transgenic mice with other experimental models for CNS inflammation such as EAE or virally-induced disease such as LCMV infection, it is apparent that many of the molecular correlates seen in individual GFAP-cytokine mice are also found in the CNS in these other models. For example, induction of the chemokine C10 and MMP-12 gene expression in CNS infiltrating macrophage/microglia and TIMP-1 gene expression in astrocytes are also prominent responses in EAE (Pagenstecher et al. 1998; Asensio et al. 1999). While the expression of IFN-regulated antiviral genes such as OAS-L2 (Sandberg et al. 1994) and the chemokine genes IP-10 and RANTES (Asensio and Campbell 1997) are highly upregulated in the brain of LCMV-infected mice. These similarities between the transgenic and the other experimental models further document the relevance of the

GFAP-cytokine transgenic mice for studying cytokine-mediated CNS inflammation.

5.4 Molecular Circuitry for Cytokine Signaling in the CNS

An issue that emerges from the preceding discussion concerns the basis for the unique repertoires of target genes that are regulated by the different astrocyte-expressed cytokines. To reconcile this, one must ultimately investigate the nature and role of the cytokine receptor-coupled signal transduction pathways that modulate gene transcriptional activity. For a significant number (30) of cytokines (e.g., IL-2, IL-3, IL-4, IL-6, IL-10, IL-12, IFN-α, -β, and -γ, and GM-CSF) and growth factors (e.g., growth hormone, prolactin, leptin, PDGF and EGF) post-receptor signaling involves specific cytoplasmic protein tyrosine kinases termed Janus kinase (JAK) and signal transducers and activators of transcription (STATs) proteins (Schindler and Strehlow 2000). There are seven known STAT proteins (STAT 1, 2, 3, 4, 5a, 5b, and 6). Binding of a cytokine such as IFN-α to its receptor results in activation and phosphorylation of two receptor-associated JAKs. These kinases then activate the cytoplasmic tails of the receptor by phosphorylating tyrosine residues. The phosphotyrosines in turn provide docking sites for STAT monomers, which bind and are subsequently phosphorylated by the JAK. The activated STAT molecules associate with each other to form either homodimers or heterodimers. This complex then translocates to the nucleus and binds to specific target DNA sequences modulating the transcription of the cytokine-regulated genes. Different cytokines activate different STAT molecules with the specificity of cytokine signaling being determined by the nature of the activated STAT dimer that forms. For example, receptor signaling by IFN-α involves the formation of STAT1/STAT2 heterodimers while for IFN-γ this involves STATI homodimers and for IL-12, STAT4 homodimers. The importance of the JAK/STAT signaling pathway in mediating cytokine actions has been established in mice with a targeted disruption of the various STAT genes (Akira 1999). For example, STAT1 knockout (KO) mice exhibit increased susceptibility to viral and other infectious diseases and cultured fibroblasts derived from these animals fail to respond to treatment with IFN (Durbin et al. 1996; Meraz et al. 1996). Similarly, STAT4 KO mice are IL-12 unresponsive and show marked impairment in their ability to mount innate and adaptive cellular immune responses (Kaplan et al. 1996; Thierfelder et al. 1996).

Members of two recently discovered families of molecules termed the suppressors of cytokine signaling (SOCS; Endo et al. 1997; Hilton et al. 1998; Starr et al. 1997), and the protein inhibitor of activated STATs (PIAS), downregulate cytokine-activated JAK/STAT signaling. The SOCS family contains at least eight members, SOCS1–7 and cytokine-inducible Src homology 2 (SH2)-domain containing protein (CIS) while the PIAS family presently contains three members, PIAS1, PIAS3, and PIASγ (Chung et al. 1997; Liu et al. 1998; Sturm et al. 2000). Expression of the SOCS and PIAS genes are upregulated by a wide range of cytokines and thus these molecules constitute a physiological negative feedback loop in the regulation of cytokine-mediated actions (Chung et al. 1997; Liu et al. 1998). Currently, the signaling-pathways affected by many of the SOCS molecules are unknown; however, studies in mice with targeted disruptions of the SOCS1, SOCS2, or SOCS3 genes reveal pivotal roles for these molecules in IFN-γ (Alexander et al. 1999; Marine et al. 1999b), growth hormone/insulin-like growth factor (IGF)-1 (Metcalf et al. 2000) or erythropoietin signaling (Marine et al. 1999a), respectively. Studies in vitro suggest that PIAS1 binds to activated STAT1 protein, preventing its binding to the GAS element and thereby suppressing IFN-γ mediated signaling (Liu et al. 1998).

Despite the critical importance of the JAK-STAT and SOCS regulatory pathways in cytokine-signaling and biologic responses, little is known concerning the expression, regulation, and role of these genes or their products in the CNS during infectious, inflammatory or other neurological diseases. Benveniste and colleagues reported that STAT1 is needed for IFN-γ induction of the class II transactivator and class II MHC genes in astrocytoma cells (Lee and Benveniste 1996) and CD40 in microglia (Nyuyen and Benveniste, in press). Early activation of STAT3 gene expression occurs in astrocytes following *N*-methyl-d-aspartate (NMDA)-induced excitotoxic cell death in vivo, suggesting a possible role for JAK/STAT signaling in gene expression leading to astrogliosis (Acarin et al. 2000).

We have recently begun to examine the nature and regulation of STAT gene expression in the CNS of the GFAP-cytokine transgenic mice (Campbell 2001). Low constitutive expression of a number of

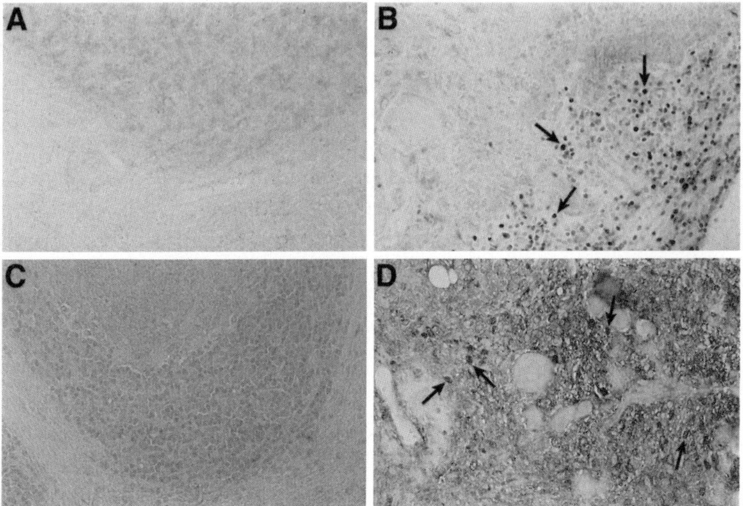

Fig. 3A–D. Immunohistochemical localization of STAT4 (**A**, **B**) and STAT1 (**C**, **D**) proteins in brain sections from wild type (**A**, **C**) and GFAP-IL-12 (**B**, **D**) mice. STAT4 protein was not detectable in brain from a wild-type mouse but is detectable in a number of mononuclear cells present within inflammatory lesions in the GFAP-IL-12 specimen. STAT1 protein was also not detectable in this specimen from a wild-type animal, however, and in contrast to STAT4, diffuse staining for STAT1 can be seen throughout the cerebellum. For both STAT4 and STAT1, nuclear localization (*arrows*) of these proteins is evident and is consistent with their activation

STAT genes including STAT1, STAT3, and STAT5 is evident in the brain of wild-type mice. There is little if any alteration in the expression of these genes in cerebrum or cerebellum from GFAP-IL-6, -IL-3, or -TNF mice. In contrast, expression of the STAT1, -2 and -3 genes is markedly increased in brain from the GFAP-IL-12 and GFAP-IFN-α transgenic mice. Additionally, the STAT4 gene is expressed in the brain of symptomatic GFAP-IL-12 mice. Thus, the expression of a number of key STAT genes in the CNS of mice is not static and can be regulated in a cytokine-dependant fashion. The level of these STAT proteins parallels the corresponding RNA transcripts in the brain from GFAP-IL-12 and GFAP-IFN-α transgenic mice. Immunohistochemical staining has revealed that while upregulated STAT1 protein can be found widely in the brain, including neurons, microglia, and astrocytes, STAT4 protein is found only in the infiltrating mononuclear cell population (Fig. 3). Nuclear localization of both STAT1 and STAT4 proteins has also been observed by us, indicating activation of these molecules. These findings show there is considerable cellular compartmentalization in the cytokine signaling pathways in the brain during inflammatory responses (J. Maier and I.L. Campbell, unpublished). This in turn infers that while the IFNs that signal via STAT1 can influence the function of a broad group of neural cells, IL-12 actions which are transduced by STAT4 are limited to cells of the immune compartment. We are currently employing mutant mice with gene deletions for STAT1 or STAT4 to facilitate a better understanding of the role of these and the other signal transduction molecules in cytokine signaling in the CNS.

As indicated above the JAK/STAT signaling pathway is subject to negative regulation by SOCS and PIAS molecules. In wild-type mice we found there is high expression of the SOCS2, -5, and PIAS1 genes, while expression of the SOCS1 and -3 RNA transcripts is quite low (J. Maier and I. Campbell, unpublished). Polizzotto and colleagues showed there are high levels of SOCS2 RNA expression by neurons in the developing and adult CNS of mice (Polizzotto et al. 2000). In the same study, lower levels of SOCS1 and SOCS3 RNA were detected throughout the brain. Studies from mice with a targeted disruption of the SOCS2 gene indicate this molecule is important in the negative regulation of growth hormone and IGF-1 signaling (Metcalf et al. 2000). Therefore, SOCS2 may have an important role in the development of neurons in the CNS and in particular direct their response to growth and differentiation factors such as IGF-1. Our studies have additionally identified high constitutive expression of the SOCS5 and PIAS1 genes in the brain; however, the significance of this is unclear. The function of SOCS5 is currently unknown while PIAS1 is known to downregulate IFN signaling. Further studies are required to determine the cellular localization and function of these signaling suppressor genes in the CNS. Overall, the expression of all these genes is unaltered in the brain of GFAP-IL-3 and -TNF mice, while SOCS3 mRNA levels are increased in the brain of GFAP-IL-6 mice. More significant increases in the expression of the SOCS1 and -3 genes is found in the brain of GFAP-IFN-α and -IL-12 mice and is largely restricted to the infiltrating mononuclear cell population (J. Maier and I.L. Campbell. unpublished).

Thus, like the expression of the STATs, the SOCS1 and -3 genes are highly regulated and exhibit cellular compartmentalization during active immune responses in the CNS.

These studies begin to define the CNS expression patterns and regulatory control of crucial components of the signaling pathways that facilitate cellular communication by a variety of cytokines in vivo. We show that the expression of key positive, i.e., STAT, and negative, i.e., SOCS, regulatory factors involved in IL-12 (STAT4), IFN-α (STAT1, STAT2, and SOCS1) and IFN-γ (STAT1, SOCS1 and -3) receptor-mediated signaling is highly regulated and compartmentalized during active immune responses in the CNS. Therefore, these findings define a molecular circuitry in the CNS that governs the cellular targets and their responses to cytokines such as IL-12 and IFN-γ.

5.5 Concluding Remarks

Understanding of the role of cytokines in regulating inflammatory processes in the living CNS has been greatly facilitated by the development of transgenic mice with cytokine gene expression targeted to astrocytes. These mice develop a spectrum of structural and functional neuropathologies that are linked to distinct neuroinflammatory responses induced by each cytokine. The neuropathological sequelae in these transgenic mice recapitulate many of those found in human neurological disease and thus support the notion that cytokines have a pathogenic role in these human disorders. The diverse neurological phenotypes induced by the astrocyte-targeted production of these different cytokines arise from their ability to differentially regulate the expression of key gene targets in specific cell types. This is accomplished through the activation of unique cytokine receptor-coupled signal transduction pathways, which for many cytokines involves the JAK/STAT signaling pathway. The expression of key positive, i.e., STAT, and negative, i.e., SOCS, regulatory factors involved in cytokine receptor-mediated signaling are themselves highly regulated and show cellular compartmentalization in the CNS. Thus, the cellular response to a given cytokine in the CNS will ultimately depend on the nature and activity of the molecular signaling circuitry together with the balance between positive and negative regulatory inputs.

Studies in the intact CNS as outlined here, will no doubt continue to further advance our understanding of the role of cytokines in the pathogenesis of CNS disease with the hope this will lead us to new targets for effective therapeutic intervention.

Acknowledgements. I am indebted to the many colleagues who have contributed over the years to the development and understanding of the transgenic mice described in this article. I would like to thank Heather Weinkauf and Mary Kies for their excellent technical assistance and Heather Kemlein for administrative assistance. The studies from my laboratory referred to in this article were funded by NIH grants MH50426, MH47680, MH62231, and NS36979. This is manuscript number 14140-NP from the Scripps Research Institute.

References

Acarin L, Gonzalez B, Castellano B (2000) STAT3 and NFkappaB activation precedes glial reactivity in the excitotoxically injured young cortex but not in corresponding distal thalamic nuclei. J Neuropathol Exp Neurol 59:151–163

Aicardi J, Goutieres F (1984) A progressive familial encephalopathy in infancy with calcifications of the basal ganglia and chronic cerebrospinal fluid lymphocytosis. Ann Neurol 15:49–54

Akira S (1999) Functional roles of STAT family proteins: lessons from knockout mice. Stem Cells 17:138–146

Akwa Y, Hassett DE, Eloranta ML, Sandberg K, Masliah E, Powell H, Whitton JL, Bloom FE, Campbell IL (1998) Transgenic expression of IFN-a in the central nervous system of mice protects against lethal neurotropic viral infection but induces inflammation and neurodegeneration. J Immunol 161:5016–5026

Alexander WS, Starr R, Fenner JE, Scott CL, Handman E, Sprigg NS, Corbin JE, Cornish AL, Darwiche R, Owczarek CM, Kay TWH, Nicola NA, Hertzog PJ, Metcalf D, Hilton DJ (1999) SOCS1 is a critical inhibitor of interferon g signaling and prevents the potentially fatal neonatal actions of this cytokine. Cell 98:597–608

Arnason B (1995) The role of cytokines in multiple sclerosis. Neurol 45 [Suppl 6]:S54–S55

Asensio VC, Campbell IL (1997) Chemokine gene expression in the brain of mice with lymphocytic choriomeningitis. J Virol 71:7832–7840

Asensio VC, Campbell IL (1999) Chemokines and their receptors in the CNS: plurifunctional mediators in diverse states. Trends Neurosci 22:504–512

Asensio VC, Lassmann S, Pagenstecher A, Steffensen SC, Henriksen SJ, Campbell IL (1999) C10 is a novel chemokine expressed in experimental inflammatory demyelinating disorders that promotes the recruitment of macrophages to the central nervous system. Am J Pathol 154:1181–1191

Black DN, Watters GV, Andermann E, Dumont C, Kabay ME, Kaplan P, Meagher-Villemure K, Michaud J, O'Gorman G, Reece E, Tsoukas C, Wainberg MA (1988) Encephalitis among Cree children in Northern Quebec. Ann Neurol 24:483–489

Butcher EC (1990) Cellular and molecular mechanisms that direct leukocyte traffic. Am J Pathol 136:3–11

Campbell IL (1998a) Transgenic mice and cytokine actions in the brain: bridging the gap between structural and functional neuropathology Brain Res Rev 26:327–336

Campbell IL (1998b) Structural and functional impact of the transgenic expression of cytokines in the CNS. Ann NY Acad Sci 840:83–96

Campbell IL (2001) Cytokine-mediated inflammation and signaling in the intact central nervous system. Prog Brain Res 132:491–508

Campbell IL, Pagenstecher A (1999) Matrix metalloproteinases and their inhibitors in the nervous system: the good, the bad and the enigmatic. Trends Neurosci 22:285–287

Campbell IL, Abraham CR, Masliah E, Kemper P, Inglis JD, Oldstone MBA, Mucke L (1993) Neurologic disease induced in transgenic mice by the cerebral overexpression of interleukin 6. Proc Natl Acad Sci USA 90:10061–10065

Campbell IL, Eddleston M, Kemper P, Oldstone MBA, Hobbs MV (1994) Activation of cerebral cytokine gene expression and its correlation with onset of reactive astrocyte and acute-phase response gene expression in scrapie. J Virol 68:2383–2387

Campbell IL, Krucker T, Steffensen S, Akwa Y, Powell HC, Lane T, Carr D, Gold LH, Henriksen SJ, Siggins GR (1999) Structural and functional neuropathology in transgenic mice with CNS expression of IFN-a. Brain Res 835:46–61

Carson MJ, Sutcliffe JG, Campbell IL (1999) Microglia stimulate naive T-cell differentiation without stimulating T-cell proliferation. J Neurosci Res 55:127–134

Chiang C-S, Stalder A, Samimi A, Campbell IL (1994) Reactive gliosis as a consequence of interleukin-6 expression in the brain. Studies in transgenic mice. Dev Neurosci 16:212–221

Chiang C-S, Powell HC, Gold L, Samimi A, Campbell IL (1996) Macrophage/microglial-mediated primary demyelination and motor disease induced by the central nervous system production of interleukin-3 in transgenic mice. J Clin Invest 97:1512–1524

Chung CD, Liao J, Liu B, Rao X, Jay P, Berta P, Shuai K (1997) Specific inhibition of Stat 3 signal transduction by PIAS3. Science 278:1803–1805

Durbin JE, Hackenmiller R, Simon MC, Levy DE (1996) Targeted disruption of the mouse Stat1 gene results in compromised innate immunity to viral disease. Cell 84:443–450

Endo TA, Masuhara M, Yokouchi M, Suzuki R, Sakamoto H, Mitsui K, Matsumoto A, Tanimura S, Ohtsubo M, Misawa H, Miyazaki T, Leonor N, Taniguchi T, Fujita T, Kanakura Y, Komiya S, Yoshimura A (1997) A new protein containing an SH2 domain that inhibits JAK kinases. Nature 387:921–924

Feuerstein G, Wang X, Barone FC (1997) Inflammatory gene expression in cerebral ischemia and trauma. Ann NY Acad Sci 825:179–193

Frendl G, Beller DI (1990) Regulation of macrophage activation by IL-3. J Immunol 144:3392–3399

Ghirnikar RS, Lee YL, Eng LF (1998) Inflammation in traumatic brain injury: role of cytokines and chemokines. Neurochem Res 23:329–340

Goetzl EJ, Banda MJ, Leppert D (1996) Matrix metalloproteinases in immunity. J Immunol 156:1–4

Griffin DE (1997) Cytokines in the brain during viral infection:clues to HIV-associated dementia. J Clin Invest 100:2948–2951

Haelens A, Wuyts A, Proost P, Struye S, Opdenakker G, van Damme J (1996) Leukocyte migration and activation by murine chemokines. Immunobiol 195:499–521

Heyser CJ, Masliah E, Samimi A, Campbell IL, Gold LH (1997) Progressive decline in avoidance learning paralleled by inflammatory neurodegeneration in transgenic mice expressing interleukin 6 in the brain. Proc Natl Acad Sci USA 94:1500–1505

Hilton DJ, Richarson RT, Alexander WS, Viney EM, Willson TA, Sprigg NS, Starr R, Nicholson SE, Metcalf D, Nicola NA (1998) Twenty proteins containing a C-terminal SOCS box form five structural classes. Proc Natl Acad Sci USA 95:114–119

Kaplan MH, Sun Y-L, Hoey T, Grusby MJ (1996) Impaired IL-12 responses and enhanced development of Th2 cells in Stat4-deficient mice. Nature 382:174–177

Lebon P, Badoual J, Ponsot G, Goutieres F, Hemeury-Cukier F, Aicardi J (1988) Intrathecal synthesis of interferon-alpha in infants with progressive familial encephalopathy. J Neurol Sci 84:201–208

Lee TT, Martin FC, Merrill JE (1993) Lymphokine induction of rat microglia multinucleated giant cell formation. Glia 8:51–61

Lee YJ, Benveniste EN (1996) Stat1 alpha expression is involved in IFN-gamma induction of the class II transactivator and class II MHC genes. J Immunol 157:1559–1568

Leist TP, Frei K, Kam-Hansen S, Zinkernagel RM, Fontana A (1988) Tumor necrosis factor a in cerebrospinal fluid during bacterial, but not viral, meningitis. J Exp Med 167:1743–1748

Lipton SA (1997) Neuropathogenesis of acquired immunodeficiency syndrome dementia. Curr Opin Neurol 10:247–253

Liu B, Liao J, Rao X, Kushner SA, Chung CD, Chang DD, Shuai K (1998) Inhibition of Stat1-mediated gene activation by PIAS1. Proc Natl Acad Sci USA 95:10626–10631

Marine J-C, McKay C, Wang D, Topham DJ, Parganas E, Nakajima H, Pendeville H, Yasukawa H, Sasaki A, Yoshimura A, Ihle JN (1999) SOCS3 is essential in the regulation of fetal liver erythropoiesis. Cell 98:617–627

Marine J-C, Topham DJ, McKay C, Wang D, Parganas E, Stravopodis D, Yoshimura A, Ihle JN (1999) SOCS1 deficiency causes a lymphocyte-dependent perinatal lethality. Cell 98:609–616

Meraz MA, White JM, Sheehan KCF, Bach EA, Rodig SJ, Dighe AS, Kaplan DH, Riley JK, Greenlund AC, Campbell D, Carver-Moore K, DuBois RN, Clark R, Aguet M, Schreiber RD (1996) Targeted disruption of the Stat1 gene in mice reveals unexpected physiologic specificity in the JAK-STAT signaling pathway. Cell 84:431–442

Metcalf D, Greenhaigh CJ, Viney E, Willson TA, Starr R, Nicola NA, Hilton DJ, Alexander WS (2000) Gigantism in mice lacking supressor of cytokine signaling-2. Nature 405:1069–1073

Nyuyen VT, Benveniste EN (20002) Involvement of STAT-1 and Ets family members in IFN-(gamma) induction of CD40 transcription in microglia/macrophages. J Biol Chem (in press)

Pagenstecher A, Stalder AK, Kincaid CL, Shapiro SD, Campbell IL (1998) Differential expression of matrix metalloproteinase and tissue inhibitor of matrix metalloproteinase genes in the mouse central nervous system in normal and inflammatory states. Am J Pathol 152:729–741

Pagenstecher A, Lassmann S, Carson MJ, Kincaid CL, Stalder AK, Campbell IL (2000) Astrocyte-targeted expression of IL-12 induces active cellular immune responses in the central nervous system and modulates experimental allergic encephalomyelitis. J Immunol 164:4481–4492

Polizzotto MN, Bartlett PF, Turnley AM (2000) Expression of "suppressor of cytokine signalling" (SOCS) genes in the developing and adult mouse central nervous system. J Comp Neurol 423:348–358

Ransohoff RM, Glabinski A, Tani M (1996) Chemokines in immune-mediated inflammation of the central nervous system. Cytokine Growth Factor Rev 7:35–46

Rogers J, Webster S, Lue LF, Brachova L, Civin WH, Emmerling M, Shivers B, Walker D, McGeer P (1996) Inflammation and Alzheimer's disease pathogenesis. Neurobiol Aging 17:681–686

Sandberg K, Eloranta M-L, Campbell IL (1994) Expression of alpha/beta interferons (IFN-a/b) and their relationship to IFN-a/b-induced genes in lymphocytic choriomeningitis. J Virol 68:7358–7366

Schall TJ, Bacon KB (1994) Chemokines, leukocyte trafficking, and inflammation. Curr Opin Immunol 6:865–873

Schindler C, Strehlow I (2000) Cytokines and STAT signaling. Adv Pharmacol 47:113–174

Stalder AK, Carson MJ, Pagenstecher A, Asensio VC, Kincaid C, Benedict M, Powell HC, Masliah E, Campbell IL (1998) Late-onset chronic inflammatory encephalopathy in immune-competent and severe combined immune-deficient (SCID) mice with astrocyte-targeted expression of tumor necrosis factor. Am J Pathol 153:767–783

Starr R, Willson TA, Viney EM, Murray LJL, Rayner JR, Jenkins BJ, Gonda TJ, Alexander WS, Metcalf D, Nicola NA, Hilton DJ (1997) A family of cytokine-inducible inhibitors of signalling. Nature 387:917–921

Steffen BJ, Breier G, Butcher EC, Schulz M, Engelhardt B (1996) ICAM-1, VCAM-1, and MAdCAM-1 are expressed on choroid plexus epithelium but not endothelium and mediate binding of lymphocytes in vitro. Am J Pathol 148:1819–1838

Sturm S, Koch M, White FA (2000) Cloning and analysis of a murine PIAS family member, PIASg, in developing skin and neurons. J Mol Neurosci 14:107–121

Suzuki Y (1999) Genes, cells and cytokines in resistance against development of toxoplasmic encephalitis. Immunobiology 201:255–271

Taub DD (1996) Chemokine-leukocyte interactions. The voodoo that they do so well. Cytokine Growth Factor Rev 7:355–376

Thierfelder WE, van Deursen JM, Yamamoto K, Tripp RA, Sarawar SR, Carson RT, Sangster MY, Vignali DA, Doherty PC, Grosveld GC, Ihle JN (1996) Requirement for Stat4 in interleukin-12-mediated responses of natural killer and T cells. Nature 382:171–174

Tolmie JL, Shillito P, Hughes-Benzie R, Stephenson JBP (1995) The Aicardi-Goutieres syndrome (familial, early onset encephalopathy with calcifications of the basal ganglia and chronic cerebrospinal fluid lymphocytosis). J Med Genet 32:881–884

Waage A, Halstensen A, Shalaby R, Brandtz P, Kierulf P, Espevik T (1989) Local production of tumor necrosis factor a, interleukin 1, and interleukin 6 in meningococcal meningitis. J Exp Med 170:1859–1867

Williams AE, van Dam A-M, Man-A-Hing WKH, Berkenbosch F, Eikelenboom P, Fraser H (1994) Cytokines, prostaglandins and lipocortin-1 are present in the brains of scrapie-infected mice. Brain Res 654:200–206

Yong VW, Krekowski CA, Forsyth PA, Bell R, Edwards DR (1998) Matrix metalloproteinases and diseases of the CNS. Trends Neurosci 21:75–80

6 BSE, Scrapie, and vCJD: Infectious Neurodegenerative Diseases

C. Riemer, D. Simon, S. Neidhold, J. Schultz, A. Schwarz, M. Baier

6.1 Introduction 85
6.2 The BSE Crisis 86
6.3 New Variant CJD 89
6.4 Pathogen/Host Interactions in Scrapie 90
6.5 Scrapie Pathogenesis in the CNS 91
6.6 Gene Expression in Scrapie-Infected Brain Tissue 92
6.7 Outlook 96
References 98

6.1 Introduction

Transmissible spongiform encephalopathies (TSEs) are infectious progressive neurodegenerative disorders. TSEs are usually associated with the appearance of an abnormal insoluble and protease-resistant form of a normal host-encoded protein, the prion protein (PrP). Structurally the disease-associated abnormal form of PrP, termed PrP^{res} or PrP^{Sc}, is characterised by a high beta-sheet content in contrast to the predominantly alpha-helical fold of normal PrP (Hope and Manson 1991; Caughey et al. 1998).

According to the prion hypothesis, the infectious agent in TSEs consists only of PrP^{res} (Cheesebro 1997; Prusiner 1998), hence these pathogens could store and maintain their "genetic" information only in the protein fold of PrP^{res}. As predicted by the prion hypothesis, mouse strains deficient for PrP are completely resistant against TSE infections (Büeler et al. 1993). Given that the pathogen may consist only of a protein encoded by the infected host, it is not surprising that humoral or cellular antigen-triggered immune responses were never detected in TSE infections (Aguzzi and Weissmann 1997). This fact combined with the absence of an infectious agent-specific nucleic acid represent major obstacles for the development of ante mortem TSE diagnostics. Up to date, the available diagnostic procedures are mainly based on the detection of PrP^{res} accumulations in infected brain tissue by immunohistochemistry, enzyme-linked immunosorbent assay (ELISA) or Western blotting.

6.2 The BSE Crisis

Besides long-known TSEs like scrapie (a naturally occurring disease in sheep and goats) along with the corresponding pathologies in humans [Creutzfeldt-Jakob disease (CJD), Gerstmann-Sträussler-Scheinker syndrome (GSS), fatal familial insomnia (FFI), and Kuru] – and TSEs in other species (see Table 1), the first confirmation of a novel TSE in cattle, the bovine spongiform encephalopathy (BSE), has attracted considerable public and scientific interest.

The origin of BSE is still subject of controversial debates. It is widely accepted that the infectious agent causing BSE was initially spread and amplified in the cattle population of the United Kingdom through the use of contaminated meat and bone meal (MBM) and the recycling of materials from infected animals for MBM production. So far in Europe, about 180,000 cattle have died of BSE, and the economical costs to overcome this agricultural crisis are immense. Measures to eradicate BSE in the UK were mainly directed against the use of MBM ingredients in animal food. The prohibition of MBM in ruminant food in 1988 was extended to the ban of MBM from food for any animals in 1996. BSE case numbers in the UK have declined since 1993, indicating that the ban of MBM from food was an adequate step to prevent de novo infections with the BSE agent.

However, in other European countries similar measures were put in place too late and were most often not accompanied by strict controls of animal food for MBM contamination. Consequently, BSE epidemics are

Table 1. Transmissible spongiform encephalopathies

Host	Encephalopathy
Human	Sporadic
	Creutzfeldt-Jakob disease (CJD)[a]
	Familial
	Gerstmann-Sträussler-Scheinker syndrome (GSS)
	Fatal familial insomnia (FFI)
	Infections
	Iatrogenic CJD – infection during surgical procedures, via Dura Mater transplantations or by contaminated human growth hormone-preparations
	New variant of CJD caused by the BSE agent (vCJD)
	Kuru – transmission of CJD through cannibalistic rituals on Papua-New-Guinea
Sheep, goat	Scrapie
Cattle	Bovine spongiform encephalopathy (BSE)
Cat	Feline spongiform encephalopathy (FSE)
Mink	Transmissible mink encephalopathy (TME)
Elk (USA, Canada)	Chronic wasting disease (CWD)

[a]About 85%–90% of all "classic" CJD cases are sporadic.

now evolving in most European countries (Table 2). Interestingly, whereas for example in France rising numbers of clinical BSE cases were diagnosed since 1993, the presence of genuine German BSE cases was only detected upon introduction of BSE screening tests of all slaughtered cattle older than 24 months in the year 2000. Anyhow, the increasing awareness of the spread of BSE has led to the total ban of MBM in animal food in Europe at the end of the year 2000 by the European Union. Taking into account the experiences made in the UK and the current development of the European BSE epidemic, it is likely that the BSE incidence will start to decline within the next 4 years in all EU-member states. However, at the same time, countries outside the EU which imported MBM or cattle from the UK or other EU countries during the last 20 years and allowed for MBM in animal food are at risk to harbour as-yet undetected BSE cases.

The possibility of BSE agent-infected sheep is also an unsettled issue. In contrast to BSE, the scrapie agent in sheep is usually considered harmless for humans. Scrapie is endemic in many European countries and elsewhere and seems to persist in sheep populations by essen-

Table 2. Confirmed BSE cases in Europe as of May 2001[a]

	DK	B	F	D	IR	LI	L	NL	P	E	CH	I	UK
1989					15 (5)								7,133
1990					14 (1)				(1)		2		14,181
1991			5		17 (2)				(1)		8		25,026
1992	(1)			(1)	18 (2)				(1)		15		36,680
1993			1		16				(3)		29		34,370
1994			4	(3)	19 (1)				12		64		23,943
1995			3		16 (1)				14		68		14,301
1996			12		73				29		45		8,013
1997		1	6	(2)	80		1	2	30		38		4,309
1998		6	18		83	2		2	106		14		3,178
1999		3	31 (1)		96			2	170		50		2,254
2000	1	9	162	7	149			2	163	2	33		1,311
2001	2	14	81	63	51			8	34	43	14	15	232

[a]Numbers in brackets indicate BSE cases among imported animals.
DK, Denmark; B, Belgium; F, France; D, Germany; IR, Ireland; LI, Liechtenstein; L, Luxembourg; NL, Netherlands; P, Portugal; E, Spain; CH, Switzerland; I, Italy; UK, United Kingdom.

tially unknown mechanisms. The possible introduction of the BSE agent into sheep populations via contaminated MBM would be hard to detect, because the symptoms of BSE and scrapie in sheep are virtually identical (Foster et al. 2001).

6.3 New Variant CJD

In 1996, cases of a new variant of CJD (vCJD) were described and it became quickly evident that vCJD, in all likelihood, represents the infection of humans with the BSE agent (Will et al. 1996; Bruce et al. 1997; Hill et al. 1997; Scott et al. 1999). These infections are probably caused by the consumption of food contaminated with the BSE agent. For the classic form of CJD, which is subdivided into sporadic, genetic, and iatrogenic cases, the annual frequency is about 1–2 patients per million per year (Brown and Gajdusek 1991). Since 1996, 100 cases of vCJD have been diagnosed and the vCJD incidence continues to rise. Epidemiological calculations predict up to 150,000 vCJD cases in the United Kingdom for the next 20–40 years. Given the most likely very long incubation times for vCJD, care has to be taken to prevent further transmission of vCJD within the human population, for example, via the blood transfusion system or through surgical procedures.

The incubation times are most likely influenced by the size of the infectious dose and by genetic factors in the human population. So far all vCJD patients were homozygous for methionine at codon 129 of the PrP gene, whereas about 60% of the population is either homozygous for valine or heterozygous (methionine/valine) at this position (Alperovitch et al. 1994). Thus, codon 129-methionine homozygosity is an apparent risk factor to develop vCJD. Recent experimental evidence suggests that other as-yet unidentified host genes in addition to PrP may determine TSE incubation times. Pessimistic predictions of the vCJD epidemic argue that the vCJD cases diagnosed so far represent only the early manifestation of the disease in the genetically most susceptible part of the population, and the huge majority of vCJD cases would remain to be seen in the years to come.

6.4 Pathogen/Host Interactions in Scrapie

Scrapie or BSE infections of mice or hamsters are the most intensively studied small animal model systems for TSEs. Although some progress has been made during the past 5 years concerning the spread of the infectious agent in the body upon oral uptake (Diringer et al. 1998), many aspects of the biology of TSEs are still in the dark. It is still unknown whether the TSE agent has to be transported out of the gut (via M-cells?) or whether neuroinvasion takes place directly at sites of the enteric nervous system ending in the submucosa of the gut (Beekes and McBride 2000).

For intraperitoneal (i.p.) infections it became evident that neuroinvasion of mice is largely dependant on the presence of mature follicular dendritic cells (FDCs). FDC-deficient mice display normal disease development after intracerebral infections but are virtually resistant against i.p. inoculation of the infectious agent (Mabbott et al. 2000). PrP^{Sc} is detectable in the spleen of infected animals, in areas where FDCs present antigen to B cells (McBride et al. 1992). The role of FDCs during oral TSE infections is unknown.

Besides nervous tissue, the only other cell type outside the CNS known to interact with the TSE agent are macrophages. In vivo depletion of spleen macrophages leads to enhanced PrP^{res} accumulation and earlier disease onset (Beringue et al. 2000). In vitro, decreasing infectivity levels were achieved upon incubation of brain homogenates from scrapie-infected mice with peritoneal macrophages (Carp and Callahan 1981, 1982). Thus, phagocytosis and degradation of PrP^{res} deposits by macrophages may represent a natural defense mechanism against TSEs. Given the participation of spleen macrophages in the clearance of PrP^{Sc}, it is interesting to note that in search for differentially expressed genes in the scrapie-infected spleen the erythroid differentiation-related factor (EDRF) was found to be downregulated (Miele et al. 2001). The expression of PrP by macrophages, which interacts with PrP^{Sc}, and the tight connection of this cell type to the erythroid lineage, may point towards a further involvement of macrophages in scrapie biology (Miele et al. 2001).

6.5 Scrapie Pathogenesis in the CNS

Infections of brain tissue by the TSE agent are characterised by a reactive gliosis and the subsequent degeneration of neuronal tissue. The activation of glial cells (for review see Rezaie and Lantos 2001), which precedes neuronal death, is likely to be caused by the deposition of large amounts of PrP^{Sc} in the brain (Williams et al. 1994, 1997; Giese et al. 1998). Experimental evidence suggests that PrP^{Sc} participates in initiating the gliosis and subsequent apoptotic and/or necrotic neuronal loss. A PrP-derived peptide (PrP106–126) activates microglia in vitro (Forloni et al. 1993; Brown et al. 1996). Furthermore, cell culture supernatants from these stimulated cells induce the proliferation of astrocytes and are toxic for neuronal cells (Brown and Kretzschmar 1997). Thus cytokines released by PrP^{Sc}-activated microglia cells may contribute to scrapie pathogenesis by enhancement and generalisation of the gliosis and via cytotoxicity for neurons. Interestingly, although neurons from PrP-knockout (KO) mice appear to be more sensitive to oxidative stress, they were reportedly almost insensitive against the toxicity released by PrP106–126-stimulated microglia (Giese et al. 1998). This would indicate that PrP^{Sc} mediated neurotoxicity requires activation of microglia in combination with the expression of PrP on neuronal cells. The function of neuronal PrP in this context remains to be determined. Moreover, the contribution of microglia activation and cytokine induction to the disease development in vivo is essentially unknown.

Microglia may recognise PrP^{Sc} via the same receptors as amyloid-beta in Alzheimer's disease (Husemann et al. 2001; Le et al. 2001). As one would expect and in good agreement with the idea of common receptors, PrP106–126 and beta-amyloid trigger virtually identical signal transduction pathways (Combs et al. 1999).

Recent reports suggest that in the presence of anti-beta-amyloid antibodies, recognition of beta-amyloid by microglia via Fc-receptors enables these phagocytes to remove beta-amyloid efficiently (Bard et al. 2000). In the absence of antibodies, amyloid would be recognised by receptors expressed on microglia which do not mediate phagocytosis but only cell activation. Immunisation against beta-amyloid in a transgenic murine AD model was shown to prevent memory loss as well as to reduce behavioural impairment together with a reduction of plaque burden (Janus et al. 2000; Morgan et al. 2000). A similar strategy to immunise mice against PrP is a promising immunotherapeutic approach. Interestingly, monoclonal antibodies (mAbs) against PrP106–126 prevent amyloid formation of the peptide, disaggregate preformed amyloid and, in consequence, neutralise peptide toxicity in vitro (Hanan et al. 2001).

Generally the development of strategies for the therapy of TSEs is severely hampered by the very limited knowledge of the underlying pathomechanisms in this group of diseases. Reports on scrapie without accumulation of PrP^{Sc} (Lasmezas et al. 1997; Raeber et al. 1997) and scrapie without the typical histopathology in the brain (Flechsig et al. 2000) illustrate that we are still far away from a consistent understanding of scrapie pathogenesis.

6.6 Gene Expression in Scrapie-Infected Brain Tissue

Previous research efforts to analyse gene or protein expression alterations in scrapie-infected brain tissue employed methods like immunohistochemistry, in situ hybridisation, subtractive hybridisation, and differential display RT-PCR (Duguid et al. 1988; Diedrich et al. 1991; Doh-ura et al. 1995; Dandoy-Dron et al. 1998). Most importantly, proinflammatory cytokines (e.g. tumour necrosis factor alpha, interleukin-1-alpha and -beta, and interleukin-6), glia activation markers (e.g. GFAP, cathepsin S, complement C1q), and indicators of cellular stress (e.g. HSP70, methallothionein II) were found to be upregulated in the scrapie-infected brain (Wietgrefe et al. 1985; Diedrich et al. 1987; Campbell et al. 1994; Kenward et al. 1994; Williams et al. 1997; Dandoy-Dron et al. 1998).

To gain further insight into altered gene expression levels in diseased brain tissue, which may be associated with or even causative for the neurodegenerative changes, we applied the suppression subtractive hybridisation (SSH) technique in combination with a differential screening approach (Diatchenko et al. 1996). Briefly, hamsters were inoculated i.p. with scrapie strain 263K (Diringer et al. 1998). Control animals were inoculated with the same volume of normal brain homogenate. Total brain RNA from the animals was isolated at the terminal stage of the disease and subjected to the SSH procedure.

Table 3. Upregulated genes in the scrapie-infected hamster brain

Identified as:	No. of clones	mRNA size (kbp)
Genes previously not known to be affected by scrapie		
B lymphocyte chemoattractant (BLC)	5	1.8
Interferon inducible protein 10 (IP-10)	4	1.6
Interferon induced Mx-protein (p78)	1	2.1
2′-5′-oligo A synthetase	1	1.9
IIGP protein	1	3
Glycoprotein-39 precursor	1	2
Vimentin	2	2.5
Aquaporin-4 (AQP-4)	3	6
Lysosomal-associated multi-trans-membrane protein (LAPTm5)	1	1.9/2.8
LIM-homeodomain 7 (Lhx7)	1	2
Complement C1q C-chain	5	1.2
Known genes		
GFAP	34	3
Transferrin	6	nd
Apolipoprotein J	4	nd
Metallothionein	2	nd
2 Microglobulin	2	nd
MHC-class I	1	nd
MHC-class II	6	nd
MHC-class II ass. invariant chain	1	nd

nd, not determined.

Twelve hundred clones generated by the SSH technique were analysed in dot-blots with forward and reverse subtracted cDNA probes from infected and uninfected hamster brain, respectively. One hundred clones displayed signal intensity differences and were further characterised by nucleotide sequencing. Sequence database searches identified eight genes previously described as upregulated in the scrapie infected brain [GFAP, transferrin, apolipoprotein J, metallothionein, β_2-microglobulin, MHC class I, MHC class II, and MHC class II-associated invariant chain]. Northern-blot analysis of the remaining clones confirmed 11 genes as differentially expressed, which were until now not known to be affected by the scrapie infection: interferon inducible protein 10 (IP-10), B lymphocyte chemoattractant (BLC), Mx-protein, (2′-5′) oligo A synthetase, IIGP protein, glycoprotein-39 precursor (gp39), vimentin,

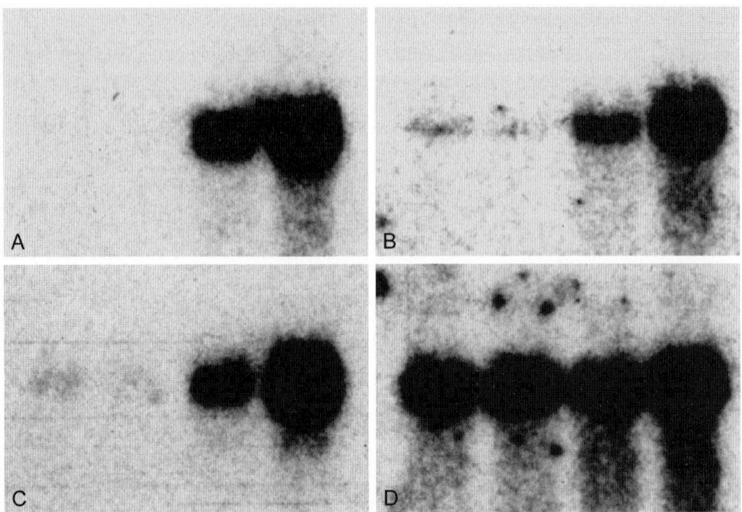

Fig. 1A–D. Northern blot confirmation of differential gene expression. Total brain RNA from two mock-infected control hamsters (*left half of each blot*) and from two scrapie-infected hamsters (*right half of each blot*) was fractionated by gel electrophoresis, blotted and hybridised with clones **A** IP-10, **B** (2′-5′) oligo A synthetase, **C** BLC, **D** beta-actin. The beta-actin probe confirms the presence of equal amounts of RNA in each lane

aquaporin-4 (AQP-4), lysosomal-associated multi-transmembrane protein (LAPTm5), the LIM-homeodomain protein 7 (Lhx7) and C1q C-chain of complement (Table 3, Fig. 1).

The levels of mRNA induction varied between the different genes from 1.5- to 6-fold as seen by Northern blotting and subsequent densitometric quantification. Only the IP-10 mRNA was virtually undetectable upon extended exposures in the uninfected brain and was highly expressed in the terminal stage of the infection (Fig. 1). In addition, expression of the chemokines IP-10 and BLC in the brain of Balb/c mice i.p. infected with the scrapie strain 139A was analysed over time (data not shown). By using a semiquantitative reverse transcriptase polymerase chain reaction (RT-PCR) method both chemokines were found to be induced at day 114 post-infection, which was about 90 days before the terminal stage of the disease.

In total, we found 19 upregulated genes (Table 3), 8 of which were previously described by other groups as differentially expressed. Among the 11 genes newly found to be affected by scrapie infection IP-10, BLC, vimentin, gp39, LAPTm5, AQP-4, and the C1q C-chain of complement are most likely expressed by activated glial cells. Increased vimentin and AQP-4 levels are indicative of an ongoing astrocytosis (Lafarga et al. 1993; Nielsen et al. 1997). Among the microgliosis markers, gp39 is seen in late stages of macrophage differentiation and is thus probably expressed by activated and differentiating microglia cells. Upregulation of the C-chain of complement C1q in stimulated microglia cells is in agreement with a similar observation for the C1q B-chain (Dandoy-Dron et al. 1998). High LAPTm5 expression levels are associated with an increased lysosomal activity of microglia cells in close vicinity to spongiform histological changes (Williams et al. 1994). This observation further substantiates the possible involvement of an aberrant activation of the microglial lysosomal system in neurodegeneration (Kopacek et al. 2000).

(2′-5′) Oligo A synthetase is most likely expressed by glial as well as neuronal cells (Asada-Kubota et al. 1997). (2′-5′) Oligo A synthetase causes activation of RNase L, which has been demonstrated to participate in apoptotic cell death via its unspecific rRNA-degrading activity (Castelli et al. 1997; Zhou et al. 1997). Hence, activation of RNase L may directly contribute to neuronal loss in scrapie. In contrast to the upregulation of IP-10 and BLC at day 114 post-infection, increasing (2′-5′) oligo A synthetase mRNA levels were first seen at day 141 by reverse transcriptase polymerase chain reaction (RT-PCR) (data not shown). Which cell types express increased levels of GTPase IIGP, the putative transcription factor Lhx7 mRNA and Mx-protein is less clear.

Mx-protein and (2′-5′) oligo A synthetase are part of the antiviral response induced by interferons. Moreover, given that the expression of IIGP, MHC class I and II, complement, beta$_2$-microglobulin, and IP-10 is also interferon (IFN) inducible, gene activation in scrapie bears all hallmarks of a typical IFN response. Interestingly, IFN-gamma was previously shown to activate microglia after stimulation with beta-amyloid (Meda et al. 1995). Furthermore, low doses of IFN-gamma but not IL-1beta, IL-2, IL-4, IL-6, or IL-12 synergistically enhance CD40 expression on microglia cells upon stimulation with beta-amyloid. Thus, IFN-gamma may contribute to the CD40/CD40L interaction-dependant activation of microglia in Alzheimer's disease as well as in scrapie (Tan et al. 1999).

Whereas in lymphoid organs, IP-10 and BLC function mainly as chemoattractants for T and B cells (Farber 1997; Gunn et al. 1998; Legler et al. 1998), respectively, the role of these chemokines in the CNS is less well understood (Horuk et al. 1997; Xia and Hyman 1999). Both chemokines are likely to be produced by activated glia cells (Majumder et al. 1996). Although elevated IP-10 levels in cerebrospinal fluid (CSF) of human immunodeficiency virus (HIV)-infected individuals was suggested to correlate with increased numbers of T cells in CSF (Kolb et al. 1999), infiltrating lymphocytes were never described in scrapie-infected brain tissue despite high levels of IP-10 and BLC expression (Riemer et al. 2000). Thus, expression of these chemokines in the CNS appears to be insufficient for the recruitment of lymphocytes into this compartment. BLC was recently described as highly upregulated in Sandhoff disease, in which a pronounced microgliosis precedes neurodegeneration (Wada et al. 2000), pointing towards its potential involvement in the activation of microglia. As IP-10- and BLC-receptor KO mice are available, scrapie infections of these KO strains should facilitate further investigations into the role of chemokine expression in the CNS.

The induction of both chemokines is seen at early stages of the scrapie infection and is sustained at high levels until the end, which indicates an involvement in disease progression. Increased levels of IP-10 and BLC mRNAs were seen in two species infected with different scrapie strains. Thus, this observation is likely to be of general relevance in TSEs and less variable than the levels of TGF-β1 and cathepsin S seen for different TSE model systems (Baker et al. 1999). In addition, IP-10 and BLC may be useful surrogate markers for the differential diagnosis of neurodegenerative diseases like TSEs.

6.7 Outlook

The development of the BSE epidemic in the UK suggests that the strictly controlled ban of MBM from animal food is sufficient to eradicate BSE. Other EU members introduced efficient measures against BSE considerably later than the UK. It may take further 4–5 years until

BSE case numbers begin to decline in these countries. The people in probably every European country were exposed to food contaminated by the BSE agent during the last 20 years. So far, 100 vCJD cases – the infection of humans with the BSE agent – have been diagnosed in the UK and for the time being the future scale of a possible vCJD epidemic in the UK and elsewhere is impossible to predict.

Presently, most urgently needed in the field of TSEs is intensified research to improve the current TSE diagnostics and to develop therapeutical strategies. Current efforts to establish ante mortem TSE diagnostic focus on very sensitive methods to detect PrP^{Sc} in body fluids or on the detection of suitable surrogate markers for the infection (Beekes et al. 1999). Whether PrP^{Sc} will ever be routinely detectable in blood remains to be seen. Available biological data from various animal model systems on the amount of TSE infectivity in blood indicate the presence of very low levels of the infectious agent during late stages of the disease (Brown et al. 1998).

Although the actual cause of the fatal neurodegeneration in TSEs is still subject of scientific debates and intense research efforts, a possible role of glia activation in this pathological process is gaining acceptance. First, experimental therapeutical trials to address, in particular, the contribution of microglia will include the use of anti-inflammatory nonsteroidal drugs as well as immunisations against PrP^{Sc} to facilitate phagocytosis and degradation of amyloid.

Finally, peptide analogs or other compounds might be developed which bind to PrP^{Sc} or normal PrP and prevent the conversion of PrP into its disease-associated isoform.

In any case, the numerous parallels between CJD, vCJD, and Alzheimer's disease suggest that, in principle at least, some of the therapeutical strategies mentioned above may be applicable to more than one form of human neurodegenerative amyloidosis.

References

Aguzzi A, Weissmann C (1997) Prion research: the next frontiers. Nature 389:795–798

Alperovitch A, Brown P, Weber T, Pocchiari M, Hofman A, Will RG (1994) The incidence of Creutzfeldt-Jakob disease in Europe. Lancet 334:918

Asada-Kubota M, Ueda T, Nakashima T, Kobayashi M, Shimada M, Takeda K, Hamada K, Maekawa S, Sokawa Y (1997) Localization of 2′,5′-oligoadenylate synthetase and the enhancement of its activity with recombinant interferon-alpha A/D in the mouse brain. Anat Embryol 195:251–257

Baker CA, Lu ZY, Zaitsev I, Manuelidis L (1999) Microglial activation varies in different models of Creutzfeldt-Jakob disease. J Virol 73:5089–5097

Bard F, Cannon C, Barbour R, Burke R-L, Games D, Grajeda H, Guido T, Hu K, Huang J, Johnson-Wood K, Khan K, Kholodenko D, Lee M, Lieberburg I, Motter R, Nguyen M, Soriano F, Vasquez N, Weiss K, Welch B, Seubert P, Schenk D, Yednock T (2000) Peripherally administered antibodies against amyloid β-peptide enter the central nervous system and reduce pathology in a mouse model of Alzheimer disease. Nat Med 6:916–919

Beekes M, McBride PA (2000) Early accumulation of pathological PrP in the enteric nervous system and gut-associated lymphoid tissue of hamsters orally infected with scrapie. Neurosci Lett 278:181–184

Beekes M, Otto M, Wiltfang J, Bahn E, Poser S, Baier M (1999) Late increase of serum S100 beta protein levels in hamsters after oral or intraperitoneal infection with scrapie agent. J Infect Dis 180:518–520

Beringue V, Demoy M, Lasmezas CI, Gouritin B, Weingarten C, Deslys JP, Andreux JP, Couvreur P, Dormont D. (2000) Role of spleen macrophages in the clearance of scrapie agent early in pathogenesis J Pathol 190:495–502

Bons N, Mestre-Frances N, Belli P, Cathala F, Gajdusek DC, Brown P (1999) Natural and experimental oral infection of nonhuman primates by bovine spongiform encephalopathy agents. Proc Natl Acad Sci USA 96:4046–4051

Brown P, Gajdusek DC (1991) The human spongiform encephalopathies: kuru, Creutzfeldt-Jakob disease, and the Gerstmann-Straussler-Scheinker syndrome. Curr Top Microbiol Immunol 172:1–20

Brown DR, Kretzschmar HA (1997) Microglia and prion disease: a review. Histol Histopathol 12:883–892

Brown DR, Schmidt B, Kretzschmar HA (1996) Role of microglia and host prion protein in neurotoxicity of a prion protein fragment. Nature 380:345–347

Brown P, Rohwer RG, Dunstan BC, MacAuley C, Gajdusek DC, Drohan WN (1998) The distribution of infectivity in blood components and plasma derivatives in experimental models of transmissible spongiform encephalopathy. Transfusion 38:810–816

Bruce ME, Will RG, Ironside JW, McConnell I, Drummond D, Suttie A, McCardle L, Chree A, Hope J, Birkett C, Cousens S, Fraser H, Bostock CJ (1997) Transmissions to mice indicate that 'new variant' CJD is caused by the BSE agent. Nature 389:498–501

Büeler H, Aguzzi A, Sailer A, Greiner RA, Autenried P, Aguet M, Weissmann C (1993) Mice devoid of PrP are resistant to scrapie. Cell 73:1339–1347

Campbell IL, Eddleston M, Kemper P, Oldstone MB, Hobbs MV (1994) Activation of cerebral cytokine gene expression and its correlation with onset of reactive astrocyte and acute-phase response gene expression in scrapie. J Virol 68:2383–2387

Carp RI, Callahan SM (1981) In vitro interaction of scrapie agent and mouse peritoneal macrophages. Intervirology 16:8–13

Carp RI, Callahan SM (1982) Effect of mouse peritoneal macrophages on scrapie infectivity during extended in vitro incubation. Intervirology 17:201–207

Castelli JC, Hassel BA, Wood KA, Li XL, Amemiya K, Dalakas MC, Torrence PF, Youle RJ (1997) A study of the interferon antiviral mechanism: apoptosis activation by the 2–5A system. J Exp Med 186:967–972

Caughey B, Raymond GJ, Bessen RA (1998) Strain-dependent differences in beta-sheet conformations of abnormal prion protein. J Biol Chem 273:32230–32235

Chesebro B (1997) Human TSE disease–viral or protein only? Nat Med 3:491–492

Combs CK, Johnson DE, Cannady SB, Lehman TM, Landreth GE (1999) Identification of microglial signal transduction pathways mediating a neurotoxic response to amyloidogenic fragments of beta-amyloid and prion proteins. J. Neurosciences 1:928–939

Dandoy-Dron F, Guillo F, Benboudjema L, Deslys JP, Lasmezas C, Dormont D, Tovey MG, Dron M (1998) Gene expression in scrapie. Cloning of a new scrapie-responsive gene and the identification of increased levels of seven other mRNA transcripts. J Biol Chem 273:7691–7697

Diatchenko L, Lau YF, Campbell AP, Chenchik A, Moqadam F, Huang B, Lukyanov S, Lukyanov K, Gurskaya N, Sverdlov ED, Siebert PD (1996) Suppression subtractive hybridization: a method for generating differentially regulated or tissue-specific cDNA probes and libraries. Proc Natl Acad Sci USA 93:6025–6030

Diedrich JF, Minnigan H, Carp RI, Whitaker JN, Race R, Frey W, Haase AT (1991) Neuropathological changes in scrapie and Alzheimer's disease are associated with increased expression of apolipoprotein E and cathepsin D in astrocytes. J Virol 65:4759–4768

Diedrich J, Wietgrefe S, Zupancic M, Staskus K, Retzel E, Haase AT, Race A (1987) The molecular pathogenesis of astrogliosis in scrapie and Alzheimer's disease. Microb Pathog 2:435–442

Diringer H, Roehmel J, Beekes M (1998) Effect of repeated oral infection of hamsters with scrapie. J Gen Virol 79:609–612

Doh-ura K, Perryman S, Race R Chesebro B (1995) Identification of differentially expressed genes in scrapie-infected mouse neuroblastoma cells. Microb Pathog 18:1–9

Duguid JR, Rohwer RG Seed B (1988) Isolation of cDNAs of scrapie-modulated RNAs by subtractive hybridization of a cDNA library. Proc Natl Acad Sci USA 85:5738–5742

Farber JM (1997) Mig and IP-10: CXC chemokines that target lymphocytes. J Leukoc Biol 61:246–57

Flechsig E, Shmerling D, 2Hegyi I, Raeber AJ, Fischer M, Cozzio A, von Mering C, Aguzzi A,Weissmann C (2000) Prion protein devoid of the octapeptide region restores susceptibility to scrapie in PrP knockout mice. Neuron 27:399–408

Forloni G, Angeretti N, Chiesa R, Monzani E, Salmona M, Bugiani O, Tagliavini F (1993) Neurotoxicity of a prion protein fragment. Nature 362:543–546

Foster JD, Parnham D, Chong A, Goldmann W, Hunter N (2001) Clinical signs, histopathology and genetics of experimental transmission of BSE and natural scrapie to sheep and goats. Vet Rec 148:165–171

Giese A, Brown DR, Groschup MH, Feldmann C, Haist I, Kretzschmar HA (1998) Role of microglia in neuronal cell death in prion disease. Brain Pathol 8:449–457

Gunn MD, Ngo VN, Ansel KM, Ekland EH, Cyster JG, Williams LT (1998) A B-cell-homing chemokine made in lymphoid follicles activates Burkitt's lymphoma receptor-1. Nature 391:799–803

Hanan E, Goren O, Eshkenazy M, Solomon B (2001) Immunomodulation of the human prion peptide 106–126 aggregation. Biochem Biophys Res Commun 280:115–120

Hill AF, Desbruslais M, Joiner S, Sidle KC, Gowland I, Collinge J (1997) The same prion strain causes vCJD and BSE. Nature 389:448–450

Hope J, Manson J (1991) The scrapie fibril protein and its cellular isoform. Curr Top Microbiol Immunol 172:57–74

Horuk RA, Martin W, Wang Z, Schweitzer L, Gerassimides A, Guo H, Lu Z, Hesselgesser J, Perez HD, Kim J, Parker J, Hadley TJ, Peiper SC (1997) Expression of chemokine receptors by subsets of neurons in the central nervous system. J Immunol 158:2882–2890

Husemann J, Loike JD, Kodama T, Silverstein SC (2001) Scavenger receptor class B type I (SR-BI) mediates adhesion of neonatal murine microglia to fibrillar beta-amyloid. J Neuroimmunol 114:142–150

Janus C, Pearson J, McLaurin J, Mathews PM, Jiang Y, Schmidt SD, Chishti MA, Horne P, Heslin D, French J, Mount HT, Nixon RA, Mercken M, Bergeron C, Fraser PE, St George-Hyslop P, Westaway D (2000) A beta peptide immunization reduces behavioural impairment and plaques in a model of Alzheimer's disease. Nature 408:979–982

Kenward N, Hope J, Landon M, Mayer RJ (1994) Expression of polyubiquitin and heat-shock protein 70 genes increases in the later stages of disease progression in scrapie-infected mouse brain. J Neurochem 62:1870–1877

Kolb SA, Sporer B, Lahrtz F, Koedel U, Pfister HW, Fontana A (1999) Identification of a T cell chemotactic factor in the cerebrospinal fluid of HIV-1-infected individuals as interferon-gamma inducible protein 10. J Neuroimmunol 93:172–181

Kopacek J, Sakaguchi S, Shigematsu K, Nishida N, Atarashi R, Nakaoke R, Moriuchi R, Niwa M, Katamine S (2000) Upregulation of the genes encoding lysosomal hydrolases, a perforin-like protein, and peroxidases in the brains of mice affected with an experimental prion disease. J Virol 74:411–417

Lafarga M, Berciano MT, Saurez I, Andres MA, Berciano J (1993) Reactive astroglia-neuron relationships in the human cerebellar cortex: a quantitative, morphological and immunocytochemical study in Creutzfeldt-Jakob disease. Int J Dev Neurosci 11:199–213

Lasmezas CI, Deslys JP, Robain O, Jaegly A, Beringue V, Peyrin JM, Dormont, D (1997) Transmission of the BSE agent to mice in the absence of detectable abnormal prion protein. Science275:402–405

Lasmezas CI, Deslys JP, Demalmay R, Adjou KT, Lamoury F, Dormont D, Robain O, Ironside J, Hauw JJ (1996) BSE transmission to macaques. Nature 381:743–744

Le Y, Yazawa H, Gong W, Yu Z, Ferrans VJ, Murphy, PM, Wang JM (2001) The neurotoxic prion peptide fragment PrP106–126 is a chemotactic agonist for the G protein-coupled receptor formyl peptide receptor-like 1. J Immunol 166:1448–1451

Legler DF, Loetscher M, Roos RS, Clark-Lewis I, Baggiolini M, Moser B (1998) B cell-attracting chemokine 1, a human CXC chemokine expressed in lymphoid tissues, selectively attracts B lymphocytes via BLR1/CXCR5. J Exp Med 187:655–660

Mabbott NA, Williams A, Farquhar CF, Pasparakis M, Kollias G, Bruce ME (2000) Tumor necrosis factor alpha-deficient, but not interleukin-6-deficient, mice resist peripheral infection with scrapie J Virol 74:3338–3344

Majumder S, Zhou LZ, Ransohoff RM (1996) Transcriptional regulation of chemokine gene expression in astrocytes. J Neurosci Res 45:758–769

McBride PA, Eikelenboom P, Kraal G, Fraser H, Bruce ME (1992) PrP protein is associated with follicular dendritic cells of spleens and lymph nodes in uninfected and scrapie-infected mice. J Pathol 1992 168:413–418

Meda, L, Cassatella MA, Szendrei GI, Otvos L, Baron P, Villalba M, Ferrari D, Rossi F (1995) Activation of microglial cells by beta-amyloid protein and interferon-gamma. Nature 374:647–560

Miele G, Manson J, Clinton M (2001). A novel erythroid-specific marker of transmissible spongiform encephalopathies. Nat Med 7:361–364

Morgan D, Diamond DM, Gottschall PE, Ugen KE, Dickey C, Hardy J, Duff K, Jantzen P, DiCarlo G, Wilcock D, Connor K, Hatcher J, Hope C, Gordon M, Arendash GW (2000) A beta peptide vaccination prevents memory loss in an animal model of Alzheimer's disease. Nature 408:982–985

Nielsen S, Nagelhus EA, Amiry-Moghaddam M, Bourque C, Agre P, Ottersen OP (1997) Specialized membrane domains for water transport in glial cells: high-resolution immunogold cytochemistry of aquaporin-4 in rat brain. J Neurosci 17:171–180

Prusiner SB (1998) Prions Proc Natl Acad Sci USA 95:13363–13383

Raeber AJ, Race RE, Brandner S, Priola SA, Sailer A, Bessen RA, Aguzzi, A (1997) Astrocyte-specific expression of hamster prion protein (PrP) renders PrP knockout mice susceptible to hamster scrapie. EMBO J 16:6057–6065

Rezaie P, Lantos PL (2001) Microglia and the pathogenesis of spongiform encephalopathies. Brain Res Rev 35:55–72

Riemer C, Queck I, Simon D, Kurth R, Baier M (2000) Identification of upregulated genes in scrapie-infected brain tissue. J Virol 74:10245–10248

Scott MR, Will R, Ironside J, Nguyen HO, Tremblay P, DeArmond SJ, Prusiner SB (1999) Compelling transgenetic evidence for transmission of bovine spongiform encephalopathy prions to humans. Proc Natl Acad Sci USA 96:15137–15142

Tan J, Town T, Paris D, Mori T, Suo Z, Crawford F, Mattson MP, Flavell RA, Mullan M (1999) Microglial activation resulting from CD40-CD40L interaction after beta-amyloid stimulation. Science 286:2352–2355

Wada R, Tifft CJ, Proia RL (2000) Microglial activation precedes acute neurodegeneration in Sandhoff disease and is suppressed by bone marrow transplantation. Proc Natl Acad Sci USA 97:10954–10959

Wietgrefe S, Zupancic M, Haase A, Chesebro B, Race R, Frey W, Rustan T, Friedman RL (1985) Cloning of a gene whose expression is increased in scrapie and in senile plaques in human brain. Science 230:1177–1179

Will RG, Ironside JW, Zeidler M, Cousens SN, Estibeiro K, Alperovitch A, Poser S, Pocchiari M, Hofman A, Smith PG (1996) A new variant of Creutzfeldt-Jakob disease in the UK. Lancet 347:921–925

Williams AE, Lawson LJ, Perry VH, Fraser H (1994) Characterization of the microglial response in murine scrapie. Neuropathol Appl Neurobiol 20:47–55

Williams A, Van Dam AM, Ritchie D, Eikelenboom P, Fraser H (1997) Immunocytochemical appearance of cytokines, prostaglandin E2 and lipocortin-1 in the CNS during the incubation period of murine scrapie correlates with progressive PrP accumulations. Brain Res 754:171–180

Xia MQ Hyman BT (1999) Chemokines/chemokine receptors in the central nervous system and Alzheimer's disease. J Neurovirol 5:32–41

Zhou A, Paranjape J, Brown TL, Nie H, Naik S, Dong B, Chang A, Trapp B, Fairchild R, Colmenares C, Silverman RH (1997) Interferon action and apoptosis are defective in mice devoid of 2′,5′-oligoadenylate-dependent RNase L. EMBO J 16:6355–6363

7 Activated Microglia in Alzheimer's Disease and Stroke

J.M. Pocock, A.C. Liddle, C. Hooper, D.L. Taylor,
C.M. Davenport, S.C. Morgan

7.1 Microglial Activation in Stroke ... 106
7.2 Microglia and Alzheimer's Disease ... 113
7.3 Common Pathways Inter-Linking Microglial Reactivity in Stroke and Alzheimer's Disease ... 120
References ... 124

Microglia, the resident macrophages of the CNS, are well established as contributing to the inflammatory response in a number of diseases including multiple sclerosis. Increasing evidence points to a role for microglia in the inflammatory cascades in Alzheimer's disease and stroke. In particular, microglial activation is apparent (Haga et al.1989; Altsteil and Sperber 1991; Morioka et al. 1991; Munoz 1991; Aisen 1997; Rupalla et al. 1998; Yrjänheikki et al. 1998; Barone and Feuerstein 1999; McGeer and McGeer 1999; Cooper et al. 2000). In normal brain tissue, microglia are present in a resting ramified state and play a neuroprotective role, limiting the spread of diffusible neurotransmitters, phagocytosing extracellular debris and destroying invading micro-organisms. Ramified microglia also secrete a number of nerve growth factors and neurotrophins which promote neuronal growth and survival (Kreutzberg 1996). Under pathological conditions, microglia proliferate and transform into active "brain macrophages". In this activated state, microglia secrete a number of inflammatory mediators including nitric oxide, glutamate, proteases, superoxide and cytokines. Thus, if microglia remain activated for a sustained period, the persistent secretion of these toxic inflammatory mediators becomes detrimental to neuronal survival.

7.1 Microglial Activation in Stroke

In models of either global or focal ischaemia the microglia response is very rapid. In rat brain following transient global ischaemia, activated microglia were observed within 30 min after reperfusion (Morioka et al. 1991; Rupalla et al. 1998). In focal ischaemia, activated microglia are observed surrounding damaged neurons in and at the periphery of an ischaemic core within a few hours of the initial insult (Dietrich et al. 1993; Rupalla et al. 1998; Kato and Walz 2000). The duration and intensity of the microglial reaction in focal ischaemia appears to vary according to the presence or absence of visible neuronal damage. Thus, in areas without neuronal damage, the microglia returned to a resting state within 7 days. However, in more vulnerable areas of the brain, such as the penumbric region, delayed neuronal death is accompanied by further activation of microglia, suggesting they were phagocytic in this situation. Microglia may remain activated for a month (Kato and Walz 2000).

Early responses to ischaemia include changes in microglial morphology; the cell body becomes hypertrophic and processes become thicker and shorter (Rupalla et al. 1998). This implies that there have been changes in the microglial cytoskeleton. In our laboratory, we have found that cultured microglia exposed to medium in which glucose was replaced by 2-deoxyglucose, underwent actin changes (Fig. 1A), lost contact readily with the cell culture dish and became apoptotic (Fig. 1C). Exposure of microglia to medium without glucose or medium in which glucose was replaced with 2-deoxyglucose did not induce inducible nitric oxide synthase (iNOS) expression whilst medium supplemented with glucose increased iNOS expression. At the same time, glucose containing medium or medium in which glucose was absent did not appear to be detrimental to microglial survival (Fig. 1) (A.C. Liddle, and J.M. Pocock, unpublished observations).

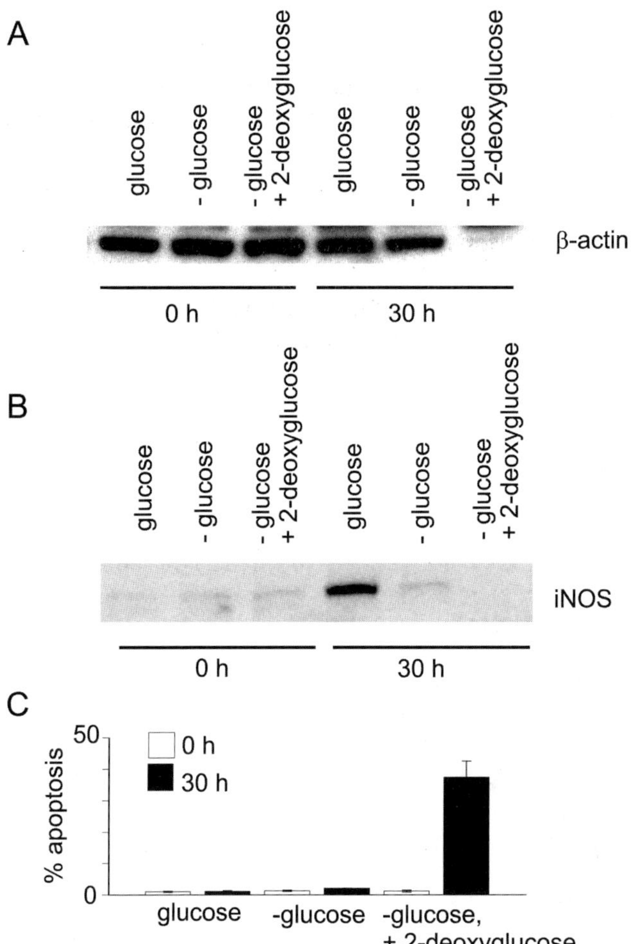

Fig. 1. A Microglia were exposed medium containing glucose, no glucose (glucose) or medium in which glucose was removed and replaced with 2-deoxyglucose (–glucose+2-deoxyglucose). Beta-actin changes were assessed 30 h later. **B** Under the same conditions as above, iNOS expression was induced in glucose-containing medium only. **C** Apoptosis in microglia exposed as above was assessed using Hoechst-33342 staining. Apoptosis was increased in microglia exposed to medium without glucose but containing 2-deoxyglucose (–glucose+2-deoxyglucose) but not in microglia exposed to medium containing glucose or with glucose removed (–glucose)

The early response of microglia at the ischaemic lesion has lead to suggestions that the primary role of the microglia is to clear the cellular debris of a lesion by phagocytosis (Bechmann and Nitsch 1997) or to promote neuronal regeneration by synapse stripping (see Streit et al. 1988). However, evidence that there is not only a primary acute and necrotic phase of neuronal death at the ischaemic core but a delayed secondary neuronal death occurring within the penumbra and that the microglial response (activation and proliferation) in areas vulnerable to secondary neuronal death precedes this delayed neuronal death, has lead to the implication of microglia in the exacerbation of this secondary damage (Rupalla et al. 1998). In vivo experiments in which the microglial and/or macrophage response to ischaemia was subdued and the recruitment of microglial and macrophages inhibited, resulted in enhanced neuronal survival (Giulian and Robertson 1990; Giulian and Vaca 1993), corroborating suggestions that microglia are involved in exacerbating neuronal loss, particularly in the delayed neuroexcitotoxic phase of ischaemia (Giulian and Vaca 1993; Kim and Ko 1998). Attenuation of microglial activation with minocycline, a tetracycline derivative, completely protected the brain against global and focal ischaemia in rodents (Tikka et al. 2001) and prevented ischaemia induced activation of microglia in a gerbil focal ischaemic model (Yrjänheikki et al. 1998). More recently, however, it has been suggested that the microglial response in ischaemia may be dependent on the level of microglial activation (Kato and Walz 2000). Those microglia, at a lesser stage of activation, which appear morphologically activated but do not express the full array of immunomolecules found on more activated phagocytic microglia may have a protective role. This is suggested by the observation that microglia have been found surrounding neurons which do not subsequently die, implying that the presence of microglia is not always pathologic (Banati and Graeber 1994; Kato et al. 1994). By secreting growth factors, these less-activated microglia may actually be neuroprotective.

The above findings indicate that microglia respond according to the severity of the ischaemic insult. Their acute sensitivity to changes in their microenvironment (Kreutzberg 1996) means that they can become moderately activated even in areas without neuronal damage and will migrate towards and undergo further activation in the ischaemic core. Their progression to a phagocytic phenotype, which may be of benefit for clearing neuronal debris, may exacerbate neuronal damage due to the ability of microglia to release a number of cytotoxic mediators.

7.1.1 Microglial Signalling in Response to Ischaemic Insults

The mechanisms by which microglia are activated by ischaemia and in particular, in areas without neuronal loss (essentially away from the ischaemic core and penumbra), are beginning to be elaborated. The induction of outward potassium currents and/or calcium transients has been found to activate microglia (Eder 1998). Ischaemic injury leads to spreading depression waves radiating from the centre of the focal injury (Irwin and Walz 1999), which may be responsible for microglial activation in areas of sub-lethal injury, or in areas with no apparent injury. The strong depolarisation induced by spreading depression in the rat brain increases the number of activated microglial cells in 3 days without an associated neuronal death (Caggiano and Kraig 1996). High extracellular K^+ and ATP release both occur during spreading depression. These stimuli can lead to an increase in intracellular calcium and the induction of calcium oscillations (Kato and Walz 2000). The frequency of these oscillations may determine which transcription factors are activated (Dolmetsch et al. 1997), thereby regulating the inflammatory response of the microglia.

Microglial migration to the lesion site may be triggered by substances released by damaged cells in the ischaemic core or the penumbra, or by a concentration gradient of substances released by spreading depression. Substances which might play a role include nitric oxide (NO), glutamate, ATP and adenosine diphosphate (ADP), and high potassium.

NO produced locally by tissue injury has been suggested to be important for microglial migration by providing a signal that microglia migrate towards and not beyond (Chen et al. 2000). NO produced at lesion sites may serve to concentrate microglia at the lesion by acting as an attractant at low concentrations and an inhibitor at higher concentrations (Magazine et al. 1996; Van Uffelen et al. 1996). The early activation and accumulation of microglia at lesion sites may be promoted by the activity of endothelial nitric oxide synthase (eNOS) rather than inducible nitric oxide synthase which takes some hours to be upregulated (Chen et al. 2000). The securing of microglia at the damaged site may at later time points be mediated by iNOS produced from reactive microglia as well as astrocytes. In order for microglia to migrate to lesion sites, changes in morphology must take place. NO can directly alter microglial morphology, through the activation of guanylate cyclase to disrupt the cytoskeleton and alter cell adhesion (Chen et al. 2000). Since changes in microglial morphology have been reported to occur as early as 30 min after an ischaemia episode, NO-induced changes may be mediated by eNOS expressed in the microglia, particularly in those microglia damaged at the lesion site, or in other glial cells (Shafer et al. 1998).

ATP, released by lysed cells at the infarct region or by spreading depression, may also activate and recruit microglia. ATP can induce chemotaxis in cultured microglia, possibly mediated by $G_{i/o}$-coupled P2Y receptors (Honda et al. 2001). Furthermore, ATP can also trigger the release of interleukin-1β, plasminogen, and tumour necrosis factor (TNF)-α (Inoue et al. 1998; Hide et al. 2000), fuelling the inflammatory response.

High levels of glutamate release following an ischaemic insult may also contribute to microglial reactivity and recruitment, and the release of substances from cells damaged by excessive glutamate-receptor stimulation may also serve as microglial attractants. Microglia have been shown to migrate to areas of neuroexcitotoxic damage induced by *N*-methyl-D-aspartate (NMDA) treatment (Heppner et al. 1998b). Furthermore, cerebral ischaemia induces tyrosine phosphorylation of PYK2 in microglial cells (Tian et al. 2000) as well as activation of p38 mitogen-activated protein kinase (p38 MAPK) (Walton et al. 1998). Glutamate can transiently activate p38 MAPK exclusively in microglia in mixed microglial/spinal-cord neuronal cultures (Tikka et al. 2001). Specific inhibitors of p38 MAPK can reduce peripheral inflammation (Underwood et al. 2000), temper microglial activation (Bhat et al. 1998) and protect the brain (gerbil) against transient forebrain ischaemia (Sugino et al. 2000). P38 MAPK in microglia also regulates iNOS and TNF-α expression (Bhat et al. 1998).

High potassium released by spreading depression could also lead to microglial activation. Stimulation of different potassium channels on microglia is implicated in their activation and proliferation (Kotecha and Schlichter 1999; Boucsein et al. 2000). Potassium channel activation can induce a respiratory burst in microglia following reoxygenation after hypoxia (Spranger et al. 1998) and increase the production of reactive oxygen species from microglia, implicating microglia in reperfusion injury after ischaemia (Spranger et al. 1998).

The proliferation of microglia in response to ischaemia may involve the protein cyclin D1. Cyclin D1 activates cyclin-dependent kinase triggering the phosphorylation of certain cellular proteins to promote the progression of cells to mitosis. Cyclin D1 mRNA is induced in response to transient global ischaemia with a similar profile of expression to that of microglial proliferation (Wiessner et al. 1996).

7.1.2 Microglia and Cell Death in Stroke

Studies aimed at teasing out the microglial response to ischaemia have shown that the vulnerability of microglia to hypoxia or hypoglycaemia is distinct (Lyons and Kettenmann 1998). Microglia were shown to be resistant to 6 h of hypoglycaemia in the presence of 2-deoxyglucose (corroborated by data in our laboratory; Liddle and Pocock 2000) and to 42 h of hypoxia in the presence of glucose. However, the combination of the two conditions lead to a loss of microglial viability. Replacement of 2-deoxyglucose with mannitol, a free radical scavenger, reduced the death, implicating a free radical component in hypoxia, hypoglycaemia-induced injury. Studies in our laboratory have shown that microglia are vulnerable to 2-deoxyglucose treatment but resistant to glucose deprivation alone, indicating that the first step in glycolysis is extremely important for microglial integrity (Liddle and Pocock 2000). Additionally, we have found that a late component of 2-deoxyglucose-induced death exhibits nuclear changes associated with apoptosis, such as nuclear condensation and chromatin fragmentation. These changes occur downstream of the mitochondrial permeability transition and after mitochondrial depolarisation. Caspase activation appears not to be required, however, since death could not be blocked by caspase inhibitors.

Recent evidence suggests that neuronal death induced by microglial activation following ischaemia may be due to the activation of Fas pathways. Fas (also known as APO-1 and CD95) is a transmembrane protein belonging to the TNF receptor family. After binding to its ligand (FasL), Fas initiates an intracellular cascade that leads to the induction of apoptosis in target cells by the activation of caspase-8 mediated pathways. In normal brain, Fas and FasL expression is low, but is upregulated following global and transient ischaemia (Matsuyama et al. 1995; Felderhoff-Mueser et al. 2000). Microglia constitutively express FasL on their cell surface and show increased expression upon activation (Spanaus et al. 1998; Vogt et al. 1998; Frigerio et al. 2000), particularly at the plasma membrane (Ciesielski-Treska et al. 2001). Plasma-membrane associated FasL can be cleaved from the cell surface by membrane associated metalloproteinases to generate the active soluble form of the ligand (sFasL) (Spanaus et al. 1998; Huang et al. 1999; Powell et al. 1999). Focal ischaemia induces early activation of metalloproteinases in rodent ischaemic models (Romanic et al. 1998; Gasche et al. 1999) and human brain infarcted tissue also displays increased metalloproteinase activity within 2 h of a stroke (Clark et al. 1997). Furthermore, microglia show increased expression of the same metalloproteinases (Anthony et al. 1997; Clark et al. 1997).

The production of reactive oxygen intermediates and ensuing oxidative stress following an ischaemic attack may also serve to increase Fas-mediated neuronal death due to microglial-derived FasL. Oxidative stress induced by hydrogen peroxide and paraquat in vitro is a highly effective potentiator of FasL expression on microglia (Vogt et al. 1998). Furthermore, graded exposure to hypoxia induces a corresponding increase in microglial FasL expression (Vogt et al. 1998). Microglia themselves are sensitive to FasL-mediated apoptosis (Boehme et al. 2000) particularly when primed with TNF-α or IFN-γ (Spanaus et al. 1998), oxidative stress or hypoxia/reoxygenation (Vogt et al. 1998). Microglia not only undergo activation but also Fas-/FasL-mediated apoptosis in response to IFN-γ or TNF-α (Badie et al. 2000; Lee et al. 2000). Thus, the increase in inflammatory cytokines or reactive oxygen species following ischaemia may serve not only to enhance microglial expression of FasL and to drive Fas-mediated apoptotic cascades in microglia, but the enhanced activity of metalloproteinases may serve to release this ligand to interact with neuronal Fas receptors.

7.2 Microglia and Alzheimer's Disease

Alzheimer's disease (AD) is a progressive neurodegenerative disease in which individuals suffer memory impairment, disorientation, confusion, impaired concentration and a decline in cognitive function. Postmortem examination of AD brain demonstrates marked atrophy of the cortical gyri, particularly in the frontal and temporal lobes of the cerebral cortex. Neuronal loss is also apparent in the locus coeruleus, raphe, nucleus basalis of Meynert and in the amygdala and hippocampus. Thus, many neurotransmitter systems are implicated in the pathology of AD including acetylcholine, noradrenaline, serotonin and dopamine. In addition to neuronal loss, characteristic neuropathological signs of disease are present in AD tissue, these include senile plaques and neurofibrillary tangles. Senile plaques accumulate in the cerebral cortex and hippocampus. The plaques consist of an amyloid core surrounded by abnormal neuronal processes, activated microglia and astrocytes. Neurofibrillary tangles are also found in the cerebral cortex and hippocampus but are composed of paired helical filaments of hyper-phosphorylated tau, an axonal microtubule-associated filament.

A prominent inflammatory response is apparent in AD (McGeer and McGeer 1999; Cooper et al. 2000), illustrated by the presence of activated and apoptotic microglia and reactive astrocytes in and around senile plaques (Haga et al. 1989), together with chronic inflammatory mediators, including components of the complement system and various cytokines (Aisen 1997; Baumel 1998). This has led to proposals that anti-inflammatory drugs might inhibit the onset and progression of AD (Stewart et al. 1997; Baumel 1998).

7.2.1 Microglial Activation by Chromogranin A

Recently, chromogranin A, a protein upregulated in the dystrophic neurites of senile plaques in AD (Munoz 1991; Yasuhara et al. 1994) has been identified as a novel and neurotoxic microglial activator (Ciesielski-Treska et al. 1998; Kingham et al. 1999; Kingham and Pocock 2000; Kingham and Pocock 2001). It is therefore plausible that CGA stimulated microglia may exacerbate neuronal damage in AD. Chromogranin A (CGA) is a polypeptide of 431–445 amino acids corresponding to a

48–52 kDa glycoprotein. It is a member of the granin family of acidic glycoproteins present in the secretory vesicles of both neuronal and endocrine cells (Hendy et al. 1995). The exact function of CGA in these cells is currently unknown, although it has been suggested that it may be involved in granule packaging. Consistent with this, CGA is present in the Golgi apparatus and in large dense-core synaptic vesicles. Furthermore, it is a precursor for a number of biologically active peptides which may act predominantly to inhibit hormone and neurotransmitter release in an autocrine or paracrine fashion (Hendy et al. 1995).

7.2.2 Signalling Cascades in Microglia Following CGA Stimulation

In AD tissue, CGA is present in the large dystrophic neurites (which may also contain amyloid precursor protein) surrounding virtually all classic senile plaques (consisting of clusters of dystrophic neurites and extracellular amyloid deposits) and to a much less extent in a sub-population of pre-amyloid plaques (consisting of neuropil deposits of beta amyloid peptide) (Munoz 1991; Yasuhara et al. 1994). This suggests that CGA may be one of the endogenous factors released from damaged neurons into brain deposits. It has been proposed that classic plaques contain a factor, absent from most pre-amyloid plaques, that can induce sprouting and enlargement of CGA-containing neurites. Furthermore, the enhanced staining of dystrophic neurites over neuropil for CGA suggests that CGA actually accumulates in dystrophic neurites. This may indicate that CGA synthesis is altered in AD. Alternatively, the accumulation of CGA in axonal swellings in AD may be due to disturbances of axonal flow and transport, as CGA is present in dense core vesicles which undergo fast axonal transport (Yasuhara et al. 1994). Blockade of retrograde transport may cause the accumulation of these large dense core vesicles in the pre-synapse of synaptic contacts present within plaques (Sahenk and Lasek 1988; Munoz 1991). Interestingly, CGA is a substrate for acetylcholinesterase, an enzyme implicated in the cholinergic hypothesis of AD (Tago et al. 1987). Furthermore, the human CGA gene is located on chromosome 14q, a region associated with the amyloid precursor protein (APP) and linked to early onset, familial AD (Schellenberg et al. 1992; Yasuhara et al. 1994).

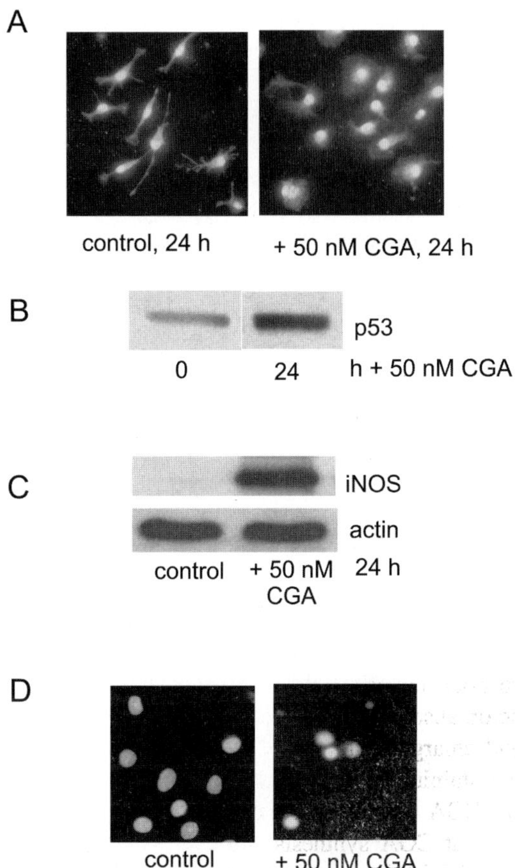

Fig. 2. **A** Control cultured rat microglia display a ramified morphology with extensive processes, whilst microglia exposed to the peptide chromogranin A (*CGA*), 50 nM for 24 h display a more reactive phenotype and an amoeboid morphology with shorter, if any, processes. **B** Microglia exposed to 50 nM CGA show increased expression of the transcription factor p53. **C** 50 nM CGA induces iNOS expression in microglia but no loss of actin. **D** Microglia exposed to 50 nM CGA show pyknotic nuclei after 24 h, following Hoechst 33342 staining

Data from our laboratory indicate that CGA induces microglia to undergo morphological changes (Fig. 2A) as well as the induction of p53 and iNOS (Fig. 2B and C) and ultimately apoptosis (Fig. 2D). In vitro, CGA-stimulated, primary cultured murine microglia follow a sequential cascade of events including iNOS expression and activation, mitochondrial depolarisation involving the mitochondrial permeability transition pore (mPT), and the activation of caspase 1, which ultimately culminates in apoptotic cell death (Kingham et al. 1999; Kingham and Pocock 2000). Increased expression of iNOS is observed after 8 h with concomitant increases in nitrite and nitrate in the culture medium. In addition, glutamate and TNF-α can be detected in the medium following iNOS induction (Taupenot et al. 1996; Kingham et al. 1999). Microglia activated with CGA also release the cysteine protease, cathepsin B (Kingham and Pocock 2001). Cathepsin B is of endosomal/lysosomal origin, and is found in particularly high abundance in activated macrophages (Lah et al. 1995; Kominami et al. 1988) where it may be important for antigen presentation (Riese and Chapman 2000). Cathepsins, including cathepsin B, can degrade extracellular matrix proteins (Buck et al. 1992), and this may aid microglia migration through the tissue. However, the degradation of extracellular matrix proteins and associated signal transduction molecules can trigger apoptosis in neighbouring cells (Bannerman et al. 1998; Levkau et al. 1998) and may have consequences for neuronal loss in chronic neurodegenerative diseases (Kingham and Pocock 2001). Whilst cathepsin B has been proposed to act as the normal intracellular secretase for the processing of amyloid precursor protein (Tagawa et al. 1991), it may also play a more direct role in the pathology of AD as senile plaques contain elevated levels of this protease (Cataldo and Nixon 1990), together with high numbers of activated microglia (Haga et al. 1989). Cathepsins D, E and B are also implicated in the processing of β-amyloid precursor protein in AD (Mackay et al. 1997) and perturbations in the activity of each cathepsin relative to others may have consequences for neurotoxicity (Frautschy et al. 1998; Ohsawa et al. 1998).

7.2.3 Microglia and Cell Death in Alzheimer's Disease

The addition of conditioned medium from CGA-stimulated microglia to cultured cerebellar granule neurons or the hippocampal cell line HT22, induces approximately 60% of the neurons to undergo apoptosis within 24 h (Kingham et al. 1999; Kingham and Pocock 2001). Cortical neurons also undergo apoptosis to a similar extent (Ciesielski-Treska et al. 1998). This neuronal death is induced ultimately by the activation of caspase-3-dependent pathways and appears to be triggered by a number of components in the conditioned medium from CGA-stimulated microglia (Kingham et al. 1999; Kingham and Pocock 2001). Interestingly, nitric oxide synthase inhibitors added to the microglia or neurons do not prevent subsequent neuronal death (Ciesielski-Treska et al. 1998; Kingham et al. 1999), indicating that NO is not a primary mediator of CGA-driven microglial neurotoxicity. Furthermore, application of TNF-α or IL-1α alone do not induce neuronal cell deathm even though TNF-α expression is increased in CGA-stimulated microglia (Taupenot et al. 1996).

Partial neuronal protection from apoptosis was achieved using antagonists to the NMDA receptor, implying that glutamate contributes to neurotoxicity, but is not solely responsible for the induction of apoptosis (Kingham et al. 1999). CGA induces the release of glutamate from microglia by reversal of the glutamate/cystine X_c transporter (Kingham et al. 1999). Glutamate may act to stimulate neuronal p38 MAP kinase pathways, since neuronal death induced by CGA-activated microglia involves the phosphorylation and activity of neuronal p38 MAPK and the release of mitochondrial cytochrome c from the neurons (Ciesielski-Treska et al. 2001). Fas-coupled apoptotic pathways may also be involved in the neuronal response to microglial neurotoxins since FasL is up-regulated in CGA-stimulated microglia (Ciesielski-Treska et al. 2001) and the generation of sFasL from microglia could induce neuronal apoptosis (Ciesielski-Treska et al. 2001). Fas expression is increased in the cerebrospinal fluid of patients with AD (Martinez et al. 2000) and is associated with glial fibrillary acidic protein (GFAP)-negative cells surrounding plaques (Nishimura et al. 1995). Whilst a full description of the role of beta-amyloid peptides in microglial activation and neuronal death is outside the scope of this paper (see Pocock and Liddle 2001), it will suffice to say that Aβ peptides are chemotactic for microglia (Lorton et al. 2000) and induce an inflammatory neurotoxic activation in microglia mediated by p38 MAPK and extracellular signal-regulated kinase (ERK) 1 and 2 (McDonald et al. 1998; Combs et al. 1999). Furthermore, amyloid proteins evoke glutamate release from microglia (Barger and Basile 2001) and induce oxidative stress (Miranda et al. 2000). Thus, it is probable that a combination of these neurotoxic inflammatory factors released from CGA-stimulated microglia act together to induce neuronal cell death through various convergent signalling cascades. Despite the growing body of in vitro evidence suggesting that inflammatory mediators play a substantial role in the neurodegeneration characteristic of AD, relatively few in vivo studies have documented elevated levels of these inflammatory factors in the CSF of AD patients. This could be attributed to difficulties incurred in obtaining age-matched control subjects or to the lack of assay sensitivity in the methods used to quantify the inflammatory mediators. Nevertheless, in vitro data provide extremely convincing evidence to suggest that inflammatory mediators released from activated microglia do indeed contribute to neurodegeneration.

In culture, impaired mitochondrial function follows inflammatory mediator release in CGA-activated microglia (Taupenot el al. 1996; Kingham et al. 1999; Kingham and Pocock 2000). Data from our laboratory indicate that the mitochondrial permeability transition is involved in CGA induced microglial apoptosis since cyclosporin A was partially protective. Mitochondrial polarity was reduced as demonstrated with the fluorescent probes JC-1 and tetra-methylrhodamine ethyl ester (Kingham and Pocock 2000). Interestingly, cytochrome c does not accumulate in the cytosol following CGA treatment. However, mitochondrial cytochrome c levels increase in two phases, one peaking at 16 h the other at 24 h. This suggests that caspase activation occurs in the mitochondrial intermembrane space or that caspase activation occurs independently of cytochrome c. In the past, apoptosis independent of cytochrome c release has been proposed to be a feature of apoptotic cell death in the absence of mitochondrial involvement. Clearly, this is not the case for CGA-induced apoptosis where a fall in the mPT can be observed at 16 h, implying a novel apoptotic cascade is in existence in CGA-stimulated microglia. In AD tissue, glia also display defective mitochondrial function, but whether the mPT is involved in this dysfunction remains unknown (de la Monte et al. 2000). However, AD brains do display extensive nicking and fragmentation of mitochondrial DNA and reduced expression of mitochondrial protein is apparent. In support of this, reduced fluorescence with mitoTracker demonstrates that a decrease in mitochondrial mass/abundance occurs in AD (de la Monte et al. 2000). Moreover, impaired mitochondrial function may render microglia more susceptible to pathological insults, thereby providing a mechanism through which environmental factors could influence the course of the disease.

Following mitochondrial dysfunction, CGA treated microglia ultimately undergo apoptotic cell death as demonstrated by the activation of caspase-1, Hoechst 33342 staining and the presence of DNA laddering. Consistent with this, CGA induces increased p53 expression in microglia (Fig. 2B, C.M. Davenport and J.M. Pocock, unpublished observations) and AD brains show increased p53 expression in both neurons and glia (microglia as well as astrocytes and oligodendrocytes) which appear to co-localise with Fas expression (de la Monte et al. 1997; Kitamura et al. 1997). Furthermore, a number of caspases (cysteine aspartases) have been identified which play a role in amyloidogenesis as well as the execution of apoptosis. Both caspase-3 and the active fragment of caspase-6 are upregulated in AD brain tissue (Le Blanc et al. 1999; Shimohama et al. 1999) and caspase-6 has been reported to directly cleave APP. Cleavage takes place at the C-terminus of APP generating a 3-kDa fragment and an Aβ-containing 6.5-kDa fragment. Pulse chase experiments suggest a precursor-product relationship between the 6-kDa fragment, intracellular Aβ and the secreted form of Aβ. These data demonstrate that, in vivo, CGA-induced caspase activity could promote neurotoxic Aβ formation, which may further exacerbate the AD phenotype.

Until recently it was believed that cell loss in AD tissue was confined to neurons. Increasing evidence indicates that microglia, as well as astrocytes and oligodendrocytes also undergo apoptosis in AD (Smale et al. 1995; de la Monte et al. 1997). Upon post-mortem examination apoptotic plaque-associated microglia can be demonstrated using an antibody specific for the C-terminal of a 32-kDa actin fragment. These fragments result from enzymatic digestion of the actin cytoskeleton which occurs during apoptosis (Yang et al. 1998). Furthermore, positive TUNEL staining for nicked DNA, as well as increased p53 and/or Fas staining give strong support for microglial apoptosis in AD (Dragunow et al. 1995; Smale et al. 1995; de la Monte et al. 1997). Frequently, Bcl-X_L-positive microglia co-localise with senile plaques, high levels of this potent apoptotic inhibitor may render microglia more resistant to cytotoxic environments (Drache et al. 1997). This suggests that microglia may exhibit self-protective mechanisms which prolong their survival under pathological conditions, which seems reasonable in light of the limited microglial replenishment mechanisms in operation in the brain. Furthermore, apoptosis of activated microglia may be an inbuilt safety measure devised to limit bystander damage to healthy neurons. Alternatively, the loss of microglia by apoptosis may limit the ability of the brain to repair itself.

The present findings describing inflammatory signalling pathways induced in CGA stimulated microglia emphasise that CGA is indeed a novel microglial activator. Since CGA is found in abundance in senile plaques and CGA-activated microglia induce neuronal apoptosis in culture; manipulation of microglial signalling cascades to limit the secretion of neurotoxic inflammatory mediators may provide a successful therapeutic strategy for the treatment of AD.

7.3 Common Pathways Inter-Linking Microglial Reactivity in Stroke and Alzheimer's Disease

Common pathways may inter-link the responses of microglia in stroke and AD (Fig. 3 and Fig. 4). Cerebrovascular disease may result in dementia and there is a complex inter-relationship between cerebrovascular disease and AD. A recent study of AD patients showed a strong correlation between the presence of brain infarcts and the occurrence of dementia and poor cognitive function (Snowdon et al. 1997). The effects of the infarcts was dependent on their location, being especially severe where subcortical and of minimal consequence if cortical. However, subcortical infarcts occurring concurrently with even minimal AD pathology resulted in severe dementia (Snowdon et al. 1997; Baumel 1998).

The brain tissue of patients with critical coronary artery disease (cCAD) exhibited the same pattern of accumulation of β-amyloid plaques as seen in AD and cCAD is a risk factor for the disease (Breteler et al. 1994; Baumel 1998). Thus, the development of a common neuropathology in these different conditions may be due to the initiation of a

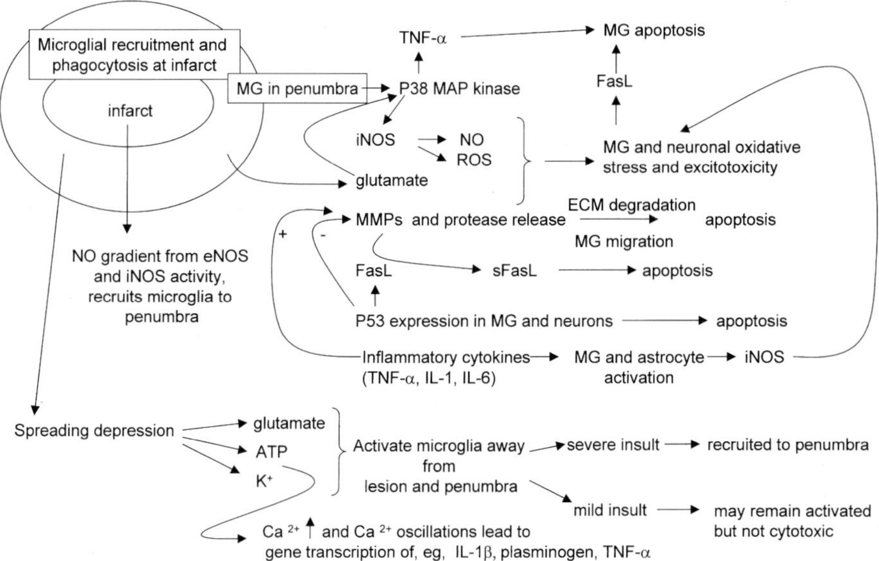

Fig. 3. Pathways activated in microglia following an ischaemic insult. Microglia at the infarct phagocytose dead and dying cells. Microglia (*MG*) outside the infarct can be recruited by a gradient of NO produced by eNOS and iNOS activity. Microglia in the penumbra become p38 MAP-kinase positive, and activation of this kinase leads to enhanced iNOS and TNF-α expression. Glutamate released from dying cells in the infarct and later penumbra can activate p38 MAP kinase on microglia and neurons to induce apoptosis. INOS activity in microglia can lead to generation of NO and reactive oxygen species (*ROS*). These, together with glutamate released from dying cells in the penumbra and possibly also from microglia, can contribute to oxidative stress of neurons and microglia. Microglia under oxidative stress show enhanced plasma membrane Fas ligand (*FasL*) expression as well as enhanced metalloproteinase (*MMPs*) and protease release. FasL can be cleaved by metalloproteinases to soluble FasL (*sFasL*) which can then induce neuronal apoptosis. Microglia in the presence of sFasL and TNF-α or IFN-γ are susceptible to FasL induced apoptosis. MMPs, and proteases released by microglia may aid microglia migration but can also degrade the extracellular matrix (*ECM*) to induce apoptosis of neurons. The expression of MMPs is negatively controlled by p53, which is also upstream of increased microglial FasL expression. Inflammatory cytokines can further exacerbate the pathways by increasing iNOS expression, increase MMP production (especially by IL-1, IL-6, TNF-α) as well as acting in concert with FasL. Spreading depression can elevate extracellular glutamate, ATP and potassium to activate microglia at sites away from the lesion and penumbra, and induce gene transcription of inflammatory mediators by increasing intracellular calcium levels

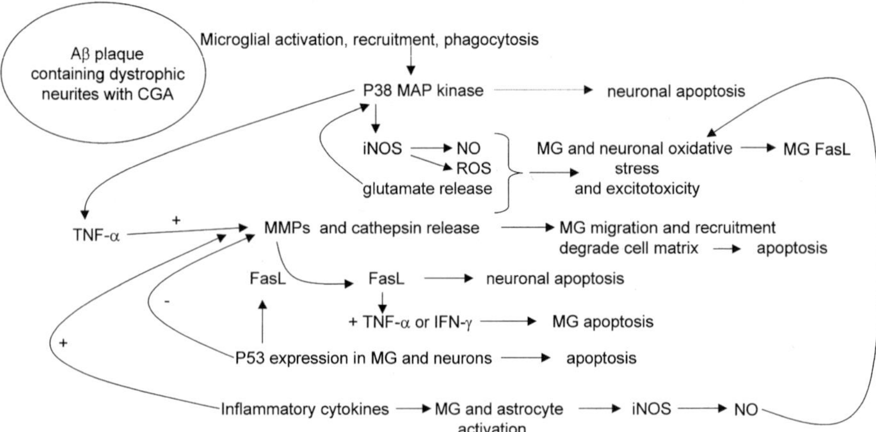

Fig. 4. Pathways activated in microglia in Alzheimer's disease. β-amyloid in the Alzheimer plaque acts as an attractant to microglia. Aβ or chromogranin-A-containing dystrophic neurites further activate microglia. P38 MAP kinase is activated and induces iNOS and TNF-α expression. INOS activation leads to the production of NO and reactive oxygen species (*ROS*) to induce microglial (*MG*) and neuronal oxidative stress. Glutamate released by CGA-stimulated MG adds to neuronal excitotoxicity. MG oxidative stress enhances the extracellular expression of Fas ligand (*FasL*) and this together with TNF-α can lead to microglial apoptosis. FasL expression is positively controlled by the tumour suppressor protein, p53, which also negatively controls metalloproteinase (*MMP*) expression. MMPs and cathepsins can aid microglial migration by degrading the extracellular matrix (*ECM*) but can also lead to apoptosis. Released MMPs can also cleave upregulated FasL from the surface of microglia to release soluble FasL (*sFasL*) to induce neuronal apoptosis. Inflammatory cytokines (particularly TNF-α, IL-1, IL-6) released from activated microglia can fuel the pathways by increasing iNOS and MMP expression

common inflammatory response which drives the development of an AD-like neuropathology (Baumel 1998).

Parallel signalling pathways may thus link the two diseases, particularly with regard to the reactions of microglia. Oxidative damage and the accumulation of free radicals is implicated in both AD (Smith et al. 1991; Smith et al. 1997; Miranda et al. 2000) and stroke (El-Kossi and Zakhary 2000), and microglia may contribute to this by the production of NO, superoxide and ultimately peroxynitrite (Colton and Gilbert

1987; Szabo and Salzman 1995). This may not only compromise neuronal function and survival but also microglial responses (Szabo and Salzman 1995; Park et al. 1999). Thus, there is increasing interest in the prophylactic use of anti-oxidant vitamins such as vitamins C, E (α-tocopherol) and β-carotene for the treatment of AD and stroke. In animal models, the free-radical scavenger vitamin E has been shown to inhibit β-amyloid induced neuronal death (Behl et al. 1992), to prevent ischaemia-induced neuronal death (Hara et al. 1990) and to significantly downregulate microglial reactivity (Heppner et al. 1998a).

Furthermore, the finding that the peptide neurotransmitter CGA is upregulated in cell swellings at the periphery of the infarct area following stroke (Yasuhara et al. 1994) and in the hippocampus following transient forebrain ischaemia in the gerbil (Marti et al. 2001) as well as in AD (Munoz 1991; Yasuhara et al. 1994), together with the discovery that CGA transforms microglia to a neurotoxic phenotype (Taupenot et al. 1996; Ciesielski-Treska et al. 1998, 2001; Kingham et al. 1999; Kingham and Pocock 2000, 2001) all point to certain common pathways in these inflammatory diseases. In addition, the increased glutamate released by microglia stimulated with CGA and Aβ peptides could contribute to the excitotoxic component of neuronal death in stroke.

Apoptotic neuronal death mediated by Fas has been reported in both AD and following stroke, and Fas and FasL expression in microglia is increased following activation (Ciesielski-Treska et al. 2001). Metalloproteases secreted by microglia (Qiu et al. 1997) could increase sFasL release from microglia in both ischaemia and AD to trigger neuronal as well as microglial apoptosis. P53 expression is increased in microglia following CGA exposure (Fig. 2) and in microglia in AD and following stroke (Kitamura et al. 1997; de la Monte et al. 1997). P53 inhibitors protect neurons against ischaemic, excitotoxic insults and β-amyloid toxicity (Culmsee et al. 2001). The presence of Aβ in the brain in patients with cCAD may serve to recruit and activate microglia via the same pathways as in AD. Furthermore, the activation of p38 MAPK pathways in both neurons and glia in stroke and AD may fuel oxidative cell death through the generation of free radicals and reactive oxygen species. Since CGA leads to the transformation of microglia to a neurotoxic phenotype and ultimately induces microglial apoptosis, this peptide may contribute to microglial over-reactivity and neuropathophysiology in stroke.

References

Aisen PS (1997) Inflammation and Alzheimer's disease; mechanisms and therapeutic strategies. Gerontology 43:143–149

Altsteil L, Sperber K (1991) Cytokines in Alzheimer's disease. Prog Neuropsychopharmacol Biol Psychiatry 15:481–495

Anthony DC, Ferguson B, Matyzak MK, Miller KM, Esiri MM, Perry VH (1997) Differential matrix metalloproteinase expression in cases of multiple sclerosis and stroke. Neuropathol Appl Neurobiol 23:406–415

Badie B, Schartner J, Vorpahl J, Preston K (2000) Interferon-gamma induces apoptosis and augments the expression of Fas and Fas ligand by microglia in vitro. Exp Neurol 162:290–296

Banati RB, Graeber MB (1994) Surveillance, intervention and cytotoxicity: is there a protective role of microglia? Dev Neurosci 16:114–127

Bannerman DD, Sathyamoorthy M, Goldblum SE (1998) Bacterial lipopolysaccharide disrupts endothelial monolayer integrity and survival signaling events through caspase cleavage of adherens junction proteins. J Biol Chem 273:35371–35380

Barger SW, Basile AS (2001) Activation of microglia by secreted amyloid precursor protein evokes release of glutamate by cystine exchange and attenuates synaptic function. J Neurochem 76:846–854

Barone FC, Feuerstein GZ (1999) Inflammatory mediators and stroke: new opportunities for novel therapeutics. J Cereb Blood Flow Metab 19:819–834

Baumel B (1998) Easing the burden of Alzheimer's disease: therapeutic options. Eur J Neurol 5 [Suppl 4]:S19-S29

Bechmann I, Nitsch R (1997) Astrocytes and microglial cells incorporate degenerating fibres following entohinal lesion: A light, confocal and electron microscopical study using a phagocytosis-dependent labelling technique. Glia 20:145–154

Behl C, Davis J, Cole GM, Schubert D (1992) Vitamin E protects nerve cells from amyloid beta protein toxicity. Biochem Biophys Res Commun 186:944–950

Bhat NR, Zhang P, Lee JC, Hogan EL (1998) Extracellular signal-regulated kinase and p38 subgroups of mitogen-activated protein kinases regulate inducible nitric oxide synthase and tumour necrosis factor-α gene expression in endotoxin-stimulated primary glial cultures. J Neurosci 18:1633–1641

Boehme SA, Lio FM, Maciejewski-Lenoir D, Bacon KB, Conlon PJ (2000) The chemokine fractalkine inhibits Fas-mediated cell death of brain microglia. J Immunol 165:397–403

Boucsein C, Kettenmann H, Nolte C (2000) Electrophysiological properties of microglial cells in normal and pathologic rat brain slices. Eur J Neurosci 12:2049–2058

Breteler MM, van Swieten JC, Bots ML, Grobbee DE, Claus JJ, van den Hout JH, van Harskamp F, Tanghe HL, de Jong PT, van Gijn J (1994) Cerebral white matter lesions, vascular risk factors, and cognitive function in a population-based study: the Rotterdam Study. Neurology 44:1246–1252

Buck MR, Karustis DG, Day NA, Honn KV, Sloane BF (1992) Degradation of extracellular matrix proteins by human cathepsin B from normal and tumour tissues. Biochem J 282:273–278

Caggiano AO, Kraig RP (1996) Eicosanoids and nitric oxide influence induction of reactive gliosis from spreading depression in microglia but not astrocytes. J Com Neurol 369:93–108

Cataldo AM, Nixon RA (1990) Enzymatically active lysosomal proteases are associated with amyloid deposits in Alzheimer brain. Proc Natl Acad Sci 87:3861–3865

Chen A, Kumar SM, Sahley CL, Muller KJ (2000) Nitric oxide influences injury-induced microglial migration and accumulation in the leech CNS. J Neurosci 20:1036–1043

Ciesielski-Treska J, Ulrich G, Taupenot L, Chasserot-Golaz S, Corti A, Aunis D, Bader M-F (1998) Chromogranin A induces a neurotoxic phenotype in brain microglial cells. J Biol Chem 273:14339–14346

Ciesielski-Treska J, Ulrich G, Ghasserot-Golaz S, Zwiller J, Revel M-O, Aunis D, Bader M-F (2001) Mechanisms underlying neuronal death induced by chromogranin A-activated microglia. J Biol Chem 276:13113–13120

Clark AW, Krekoski CA, Bou SS, Chapman KR, Edwards DR (1997) Increased gelatinase A (MMP-2) and gelatinase B (MMP-9) activities in human brain after focal ischaemia. Neurosci Lett 238:53–56

Combs CK, Johnson DE, Cannady SB, Lehman TM, Landreth GE (1999) Identification of microglial signal transduction pathways mediating a neurotoxic response to amyloidogenic fragments of β-amyloid and prion proteins. J Neurosci 19:928–939

Cooper NR, Kalaria RN, McGeer PL, Rogers J (2000) Key issues in Alzheimer's disease inflammation. Neurobiol Aging 21:451–453

Culmsee C, Zhu X, Yu Q-S, Chan SL, Camandola S, Guo Z, Greig NH, Mattson MP (2001) A synthetic inhibitor of p53 protects neurons against death induced by ischaemic and excitotoxic insults and amyloid beta peptide. J Neurochem 77:220–228

de la Monte SM, Sohn YK, Wands JR (1997) Correlates of p53- and Fas (CD95)-mediated apoptosis in Alzheimer's disease. J Neurological Sci 152:73–83

de la Monte SM, Luong T, Neely TR, Robinson D, Wands JR (2000) Mitochondrial DNA damage as a mechanism of cell loss in Alzhiemer's disease. Lab Invest 80:1323–1335

Dietrich WD, Busto R, Alonso O, Globus MYT, Ginsberg MD (1993) Intra-ischaemic but not post-ischaemic brain hyopthermia protects chronically following global forebrain ischaemia in rats. J Cereb Blood Flow Metab 13:541–549

Dolmetsch RE, Lewis RS, Goodnow CC, Healy JI (1997) Differential activation of transcription factors induced by calcium response amplitude and duration. Nature 386:855–858

Drach B, Diehl GE, Beyreuther K, Perlmutter LS, Konig G (1997) BCL-X_L-specific antibody labels activated microglia associated with Alzheimer's disease and other pathological states. J Neurosci Res 47:98–108

Dragunow M, Faull RL, Lawlor P, Beilharz EJ, Singleton K, Walker EB, Mee E (1995) In situ evidence for DNA fragmentation in Huntington's disease striatum and Alzheimer's disease temporal lobes. Neuroreport 6:1053–1057

Eder C (1998) Ion channels in microglia (brain macrophages). Am J Physiol 275:C327-C342

El-Kossi MM, Zakhary MM (2000) Oxidative stress in the context of acute cerebrovascular stroke. Stroke 31:1889–1892

Felderhoff-Mueser U, Taylor DL, Greenwood K, Kozma M, Stibenz D, Joashi UC, Edwards AD, Mehmet H (2000) Ras/CD95/APO-1 can function as a death receptor for neuronal cells in vitro and in vivo and is upregulated following cerebral hypoxic-ischaemic injury to the developing rat brain. Brain Pathol 10:17–29

Frautschy SA, Horn DL, Sigel JJ, Harris-White ME, Medoza JJ, Yang FS, Saido TC, Cole GM (1998) Protease inhibitor coinfusion with amyloid beta-protein results in enhanced deposition and toxicity in rat brain. J Neurosci 18:831108321

Frigerio S, Silei V, Ciusani E, Massa G, Lauro GM, Salmaggi A (2000) Modulation of fas-ligand (FasL) on human microglial cells: an in vitro study. J Neuroimmunol 105:109–114

Gasche Y, Fujimura M, Morita-Fukimura Y, Copin JC, Kawase M, Massengale J, Chan PH (1999) Early appearance of activated matrix metalloproteinase-9 after focal cerebral ischaemia in mice: a possible role in blood-brain barrier dysfunction. J Cereb Blood Flow Metab 19:1020–1028

Giulian D, Vaca K (1993) Inflammatory glia mediate delayed neuronal damage after ischaemia in the central nervous system. Stroke [Suppl] 12:I84–I90

Haga S, Akai K, Ishii T (1989) Demonstration of microglial cells in and around senile plaques in the Alzheimer brain. Acta Neuropath (Berl) 77:569–575

Hara H, Kato H, Kogure K (1990) Protective effect of alpha-tocopherol on ischaemic neuronal damage in the gerbil hoppocampus. Brain Res 510:335–338

Hendy GN, Bevan S, Mattei MG, Mouland AJ (1995) Chromogranin A. Clin Invest Med. 18:47–65

Heppner FL, Roth K, Nitsch R, Hailer NP (1998a) Vitamin E induces ramification and down-regulation of adhesion molecules in cultured microglial cells. Glia 22:180–188

Heppner FL, Skutella T, Hailer NP, Haas D, Nitsch R (1998b) Activated microglial cells migrate towards sites of excitotoxic neuronal injury inside organotypic hippocampal slice cultures. 10:3284–3290

Hide I, Tanaka M, Inoue A, Inoue K, Kohsaka S, Nakata Y (2000) Extracellular ATP triggers TNF-α release from rat microglia. J Neurochem 75:965–972

Honda S, Sasaki Y, Ohsawa K, Imai Y, Nakamura Y, Inoue K, Kohsaka S (2001) Extracellular ATP or ADP induce chemotaxis of cultured microglia through $G_{i/o}$-coupled P2Y receptors. J Neurosci 21:1975–1982

Huang DC, Hahne M, Schroeter M, Frei K, Fontana A, Villunger A, Newton K, Tschopp J, Strasser A (1999) Activation of Fas by FasL induces apoptosis by a mechanism that cannot be blocked by Bcl-2 or Bcl-x(L). Proc Natl Acad Sci USA 96(26):14871–14876

Inoue K, Nakajima K, Morimoto T, Kikuchi Y, Koizumi S, Illes P, Kohsaka S (1998) ATP stimulation of Ca^{2+}-dependent plasminogen release from cultured microglia. Br J Pharmacol 123:1304–1310

Irwin A, Walz W (1999) Spreading depression waves as mediators of secondary injury and of protective mechanisms. In: Walz W (ed) Cerebral Ischaemia. Humana Press, Totowa, pp 35–44

Kato H, Walz W (2000) The initiation of the microglial response. Brain Pathol 10:137–143

Kato H, Kogure K, Araki T, Itoyama Y (1994) Astroglial and microglial reactions in the gerbil hippocampus with induced ischaemic tolerance. Brain Res 664:101–107

Kim WK, Ko KH (1998) Potentiation of *N*-methyl-d-aspartate-mediated neurotoxicity by immunostimulated murine microglia. J Neurosci Res 54:17–26

Kingham PJ, Pocock JM (2000) Microglial apoptosis induced by chromogranin A is mediated by mitochondrial depolarisation and the permeability transition but not by cytochrome c release. J Neurochem 74:1452–1462

Kingham PJ, Pocock JM (2001) Microglial secreted cathepsin B induces neuronal apoptosis. J Neurochem 76:1475–1584

Kingham PJ, Cuzner ML, Pocock JM (1999) Apoptotic pathways mobilized in microglia and neurons as a consequence of chromogranin A-induced microglial activation. J Neurochem 73:538–547

Kitamura Y, Shimohama S, Kamoshima W, Matsuoka Y, Nomura Y, Taniguchi T (1997)Changes of p53 in the brains of patients with Alzheimer's disease. Biochem Biophys Res Commun 232(2):418–421

Kominami E, Tsukahara T, Hara K, Katunuma N (1988) Biosynthesis and processing of lysosomal cysteine proteinases in rat macrophages. FEBS Lett 231:225–228

Kotecha SA, Schlichter LC (1999) A Kv1.5 to Kv1.3 switch in endogenous hippocampal microglia and a role in proliferation. J Neurosci 19:10680–10693

Kreutzberg GW (1996) Microglia: a sensor for pathological events in the CNS. Trends Neurosci 19:312–318

Lah TT, Hawley M, Rock K L, Goldberg A L (1995) Gamma-interferon causes a selective induction of the lysosomal proteases, cathepsin-B and cathepsin-L, in macrophages. FEBS Lett 363:85–89

Le Blanc A, Liu H, Goodyer C, Bergeron C, Hammond J (1999) Caspase-6 role in apoptosis of human neurons, amyloidogenesis and Alzheimer's disease. J Biol Chem 274:23426–23436

Lee SJ, Zhou T, Choi C, Wang Z, Benveniste EN (2000) Differential regulation and function of Fas expression on glial cells. J Immunol 164:1277–1285

Levkau B, Herren B, Koyama H, Ross R, Raines E W (1998) Caspase-mediated cleavage of focal adhesion kinase pp125(FAK) and disassembly of focal adhesions in human endothelial cell apoptosis. J Exp Med 187:579–586

Liddle AC, Pocock JM (2000) Microglial signalling in ischaemia. Soc Neurosci Abs 767.8, vol 26, New Orleans

Lorton D, Schaller J, Lala A, De Nardin E (2000) Chemotactic-like receptors and Aβ peptide induced responses in Alzheimer's disease. Neurobiol Aging 21:463–473

Lyons SA, Kettenmann H (1998) Oligodendrocytes and microglia are selectively vulnerable to combined hypoxia and hypoglycaemia injury in vitro. J Cereb Blood Flow Metab 18:521–530

Mackay EA, Ehrhard A, Moniatte M, Guenet C, Tardif C, Tarnus C, Sorokine O, Heintzelmann B, Nay C, Remy JM, Higaki J, Van-Dorsselaer A, Wagner J, Danzin C, Mamont P (1997) A possible role for cathepsins D, E and B in the processing of beta-amyloid precursor protein in Alzheimer's disease. Eur J Biochem 244:414–425

Magazine HI, Liu Y, Bilfinger TV, Fricchione GL, Stefano GB (1996) Morphine-induced conformational changes in human monocytes, granulocytes and endothelial cells and in invertebrate immunocytes and microglia are mediated by nitric oxide. J Immunol 156:4845–4850

Marti E, Ferrer I, Blasi J (2001) Differential regulation of chromogranin A, chromogranin B and secretoneurin protein expression after transient forebrain ishchaemia in the gerbil. Acta Neuropathol (Berl) 101:159–166

Martinez M, Fernandez-Vivancos E, Frank A, De la Fuente M, Hernanz A (2000) Increased cerebrospinal fluid Fas (Apo-1) levels in Alzheimer's disease. Relationship with IL-6 concentrations. Brain Res 869:216–219

Matsuyama T, Hata R, Yamamoto Y, Tagaya M, Akita H, Uno H, Wanaka A, Furuyama J, Sugita M (1995) Localisation of Fas antigen mRNA induced in postischaemic murine forebrain by in situ hybridization. Brain Res Mol Brain Res 34:166–172

McDonald DR, Bamberger ME, Combs CK, Landreth GE (1998) β-amyloid fibrils activate parallel mitogen-activated protein kinase pathways in microglia and THP1 monocytes. J Neurosci 18:4451–4460

McGeer EG, McGeer PL (1999) Brain inflammation in Alzheimer disease and the therapeutic implications. Curr Pharm Res 5:821–838

Miranda S, Opazo C, Larrondo LF, Muñoz FJ, Ryiz F, Leighton F, Inestrosa NC (2000) The role of oxidative stress in the toxicity induced by amyloid β-peptide in Alzheimer's disease. Prog Neurobiol 62:633–648

Morioka T, Kalehua AN, Streit WJ (1991) The microglial reaction in the rat dorsal hippocampus following transient forebrain ischaemia. J Cereb Blood Flow Metab 11:966–973

Munoz DG (1991) Chromogranin A-like immunoreactive neurites are major constitutents of senile plaques. Lab Invest 64:826–832

Nishimura T, Akiyama H, Yonehara S, Kondo H, Ikeda K, Kato M, Iseki E, Kosaka K (1995) Fas antigen expression in brains with Alzheimer type dementia. Brain Res 695:137–145

Ohsawa Y, Isahara K, Kanamori S, Shibata M, Kametaka S, Gotow T, Watanabe T, Kominami E, Uchiyama Y (1998) An ultrastructural and immunohistochemical study of PC12 cells during apoptosis induced by serum deprivation with special reference to autophagy and lysosomal cathepsins. Arch Histol Cytol 61:395–403

Park LCH, Zhang H, Sheu KFR, Calingasan NY, Kristal BS, Lindsay JG, Gibson GE (1999) Metabolic impairment induces oxidative stress, compromises inflammatory responses and inactivates a key mitochondrial enzyme in microglia. J Neurochem 72:1948–1958

Pocock JM, Liddle AC (2001) Microglial signalling cascades in neurodegenerative disease. In: Castellano Lopez B, Nieto-Sampedro M (eds) Progress in brain research, vol 132. Elsevier, The Netherlands, pp 555–565

Qiu QW, Ye Z, Kholodenko D, Seubert P, Selkoe DJ (1997) Degradation of amyloid β-protein by a metalloprotease secreted by microglia and other neural and non-neural cells. J Biol Chem 272:6641–6646

Riese RJ, Chapman HA (2000) Cathepsins and compartmentalization in antigen presentation. Cur Opinion Immunol 12:107–113

Romanic AM, White RF, Arleth AJ, Ohlstein EH, Barone FC (1998) Matrix metalloproteinase expression increases after cerebral focal ischaemia in rats: inhibition of matrix metalloproteinase-9- reduces infarct size. Stroke 29:1020–30

Rupalla K, Allegrini PR, Sauer D, Weissner C (1998) Time course of microglia activation and apoptosis in various brain regions after permanent focal cerebral ischaemia in mice. Acta Neuropathol 96:172–178

Sahenk Z, Lasek RJ (1988) Inhibition of proteolysis blocks anterograde-retrograde conversion of axonally transported vesicles. Brain Res 460:199–203

Schellenberg GD, Bird TD, Wijsman EM, Orr HT, Anderson L, Nemens E, White JA, Bonnycastle L, Weber J, Alonso ME, Potter H, Heston LL, Martin GM (1992) Genetic linkage evidence for a familial Alzheimer's disease locus on chromosome 14. Science 258:668–671

Shafer OT, Chen A, Kumar SM, Muller KJ, Sahley CL (1998) Injury induced expression of endothelial nitric oxide synthase by glial and microglial cells in the leech central nervous system within minutes after injury. Proc R Soc Lond B Biol Sci 265:2171–2175

Shimohama S, Tanino H, Fujimoto S (1999) Changes in caspase expression in Alzheimer's disease: comparison with development and aging. Biochem Biophys Res Com 256:381–384

Smale G, Nichols R, Brady DR, Finch CE, Horton Jr WE (1995) Evidence for apoptotic cell death in Alzheimer's disease. Exp Neurol 133:225–230

Smith CD, Carney JM, Starke-Reed PE, Oliver CN, Stadtman ER, Floyd RA, Markesbery WR (1991) Excess brain protein oxidation and enzyme dysfunction in normal aging and Alzheimer disease. Proc Natl Acad Sci USA 88:10540–10543

Smith MA, Richey Harrus OL, Sayre LM, Beckman JS, Perry G (1997) Widespread peroxynitrite-mediated damage in Alzheimer's disease. J Neurosci 17:2653–2657

Snowdon DA, Greiner LH, Mortimer JA, Riley KP, Greiner PA, Markesbery WR (1997) Brain infarction and the clinical expression of Alzheimer disease. JAMA 277:813–817

Spanaus KS, Schlapbach R, Fontana A (1998) TNF-alpha and IFN-gamma render microglia sensitive to Fas ligand-induced apoptosis by induction of Fas expression and down-regulation of Bcl-2 and Bcl-xL. Eur J Immunol 28:4398–4408

Spranger M, Kiprianova I, Krempien S, Schwab S (1998) Reoxygenation increases the release of reactive oxygen intermediates in murine microglia. J Cereb Blood Flow Metab 18:670–674

Stewart WF, Kawas C, Corrado M, Metter EJ (1997) Risk of Alzheimer's disease and the duration of NSAID use. Neurology 48:626–632

Streit WJ, Graeber MB, Kreutzberg GW (1988) Functional plasticity of microglia: a review. Glia 1:301–307

Sugino T, Nozaki K, Takagi Y, Hattori I, Hashimoto N, Moriguchi T, Nishida E (2000) Activation of mitogen-activated protein kinases after transient forebrain ischaemia in gerbil hippocampus. J Neurosci 20:4506–4514

Szabo C, Salzman AL (1995) Endogenous peroxynitrite is involved in the inhibition of mitochondrial respiration in immunostimulated J774.2 macrophages. Biochem Biophys Res Commun 290:739–743

Tagawa K, Kunishita T, Maruyama K, Yoshikawa K, Hominami E, Tsuchiya T, Suzuki K, Tabira T, Sugita H, Isiura S (1991) Alzheimer's disease amyloid B-clipping enzyme (APP secretase): identification, purification and characterization of the enzyme. Biochem Biophys Res Commun 177:377–387

Tago H, McGeer PL, McGeer EG (1987) Acetylcholinesterase fibres and the development of senile plaques. Brain Res 507:363–369

Taupenot L, Ciesielski-Treska J, Ulrich G, Chasserot-Golaz S, Aunis D, Bader M-F (1996) Chromogranin A triggers a pheotypic transformation and the generation of nitric oxide in brain microglial cells. Neuroscience 72:377–389

Tian D, Litvak V, Lev S (2000) Cerebral ischaemia and seizures induce tyrosine phosphorylation of PYK2 in neurons and microglial cells. J Neurosci 20:6478–6487

Tikka T, Fiebich BL, Goldsteins G, Keinänen R, Koistinaho J (2001) Minocycline, a tetracycline derivative, is neuroprotective against excitotoxicity by inhibiting activation and proliferation of microglia. J Neurosci 21:2580–2588

Underwood DC, Osborn RR, Bochnowicz S, Webb EF, Rieman DJ, Lee JC, Romanic AM, Adams JL, Hay DW, Griswold DE. (2000) SB 239063, a p38 MAPK inhibitor, reduces neutrophilia, inflammatory cytokines, MMP-9, and fibrosis in lung. Am J Physiol Lung Cell Mol Physiol 279(5):895–902

Van Uffelen BE, de Koster BM, Van den Broek PJA, Van Steveninck J, Elferink JGR (1996) Modulation of neurtrophil migration by exogenous gaseous nitric oxide. J Leukocyte Biol 60:94–100

Vogt M, Bauer MK, Ferrari D, Schulze-Osthoff K (1998) Oxidative stress and hypoxia/reoxygenation trigger CD95 (APO-1/Fas) ligand expression in microglial cells. FEBS Lett 429:67–72

Walton KM, Dirocco R, Bartlett BA, Koury E, Marcy VR, Jarvis B, Schaefer EM, Bhat RV (1998) Activation of p38 MAPK in microglia after ischaemia. J Neurochem 70:1764–1767

Weissner C, Brink I, Lorenz P, Neumann-Haetelin T, Vogel P, Yamashita K (1996) Cyclin D1 messenger RNA is induced in microglia rather than neurons following transient forebrain ischaemia. Neurosci 72:947–958

Yang F, Sun X, Beech W, Teter B, Wu S, Sigel J, Vinters HV, Frautschy SA, Cole GM (1998) Antibody to caspase-cleaved actin detects apoptosis in differentiated neuroblastoma and plaque-associated neurons and microglia in Alzheimer's disease. Am J Pathol 152:379–389

Yasuhara O, Kawamata T, Aimi Y, McGeer EG, McGeer PL (1994) Expression of chromogranin A in lesions in the central nervous system from patients with neurological diseases. Neurosci Lett 170:13–16

Yrjänheikki J, Keinänen R, Pellikka M, Hökfelt T, Koistinaho J (1998) Tetracyclines inhibit microglial activation and are neuroprotective in global brain ischaemia. Proc Natl Acad Sci USA 95:15769–15774

8 Inflammation in Stroke – A Potential Target for Neuroprotection?

J. Priller, U. Dirnagl

8.1	Introduction	133
8.2	Contribution of Circulating Leukocytes to Post-Ischemic Damage	134
8.3	Molecular and Cellular Mediators of Post-Ischemic Inflammatory Responses	137
8.4	Potential Benefits of Inflammation	144
8.5	Therapeutic Perspective and Conclusion	145
References		146

8.1 Introduction

Despite of, or in fact due to, the progress in understanding the basic pathophysiology of focal cerebral ischemia ("stroke") over the past several years, clinicians and researchers alike have been disappointed by the outcome of clinical trials aimed at decreasing damage after cerebral ischemia. Currently, only thrombolytic therapy initiated within a narrow time frame of hours after the onset of ischemia has become an accepted therapeutic option (The National Institute of Neurological Disorders and Stroke rt-PA Stroke Study Group 1995). While rapid restoration of blood flow to the ischemic territories is vital, the brain vasculature unfortunately is more than a set of conduits that require "deplugging" after stroke. In fact, active changes take place at the blood-endothelial interface within minutes of stagnant blood supply, leading to an inflammatory reaction and to the transition from ischemic to inflammatory injury. Under normal circumstances, endothelial cells present an anti-thrombotic and anti-inflammatory surface to the blood. However, the release of chemokines and cytokines from brain-derived and circulating cells after ischemia triggers a cascade of potentially damaging events. The up-regulation of adhesion molecules and the subsequent attachment of peripheral blood leukocytes to endothelium may impair the restoration of blood flow in the cerebral microvasculature, damage the integrity of the blood–brain barrier and induce a prothrombotic state of the endothelium. Vascular matrix dissolution may result in hemorrhagic transformation and neuron injury, but may also be required for neovascularization. Direct neurotoxicity may derive from the generation of free radicals and the release of proteases by invading leukocytes and resident glial cells. On the other hand, pro-inflammatory cytokines may also stimulate the synthesis of growth factors required for post-injury plasticity, and activated T cells can protect neurons from secondary damage. Today, the molecular and cellular mediators of post-ischemic inflammatory processes are only just beginning to be elucidated. This review attempts to summarize current concepts of the role of inflammation following cerebral ischemia. We shall conclude that while many effects of post-ischemic inflammation are detrimental to the brain, beneficial functions also exist and have to be considered when targeting inflammation therapeutically in stroke patients.

8.2 Contribution of Circulating Leukocytes to Post-Ischemic Damage

While the causative role of neutrophils in reperfusion injury remains to be established, three lines of evidence suggest that leukocytes are involved in secondary brain damage after ischemia. First, there is a striking correlation between the time of neutrophil accumulation and the expansion of cerebral damage. Second, elevated white blood cell counts are associated with an increased risk of ischemic disease, while neutropenia leads to reduced infarct size. Third, treatment preventing neutrophil adhesion is beneficial in transient ischemia. It should be noted that the experimental evidence in support of a deleterious role of leukocytes comes mostly from studies of transient ischemia/reperfusion. Permanent ischemia also elicits an inflammatory response, but the accumulation of neutrophils is significantly lower (Barone et al. 1991).

Intense leukocyte infiltration of the brain parenchyma was found 48–72 h after the stroke in a human autopsy study (Chuaqui and Tapia 1993). Neutrophil pleocytosis in the cerebrospinal fluid (CSF) of stroke patients peaks around the same time (Sörnäs et al. 1972). Using ^{111}indium-labeled neutrophils and mononuclear cells, Pozzilli et al. observed a delayed leukocyte appearance between days 2 and 14 after symptom onset in stroke patients (Pozzilli et al. 1985a). Injection of radiolabeled autologous neutrophils in a model of air embolism-induced cerebral ischemia in dogs revealed progressive neutrophil accumulation in the brain regions with low blood flow and a significant correlation with the severity of ischemia (Kochanek et al. 1987). In baboons, del Zoppo et al. used 3 h of middle cerebral artery occlusion (MCAO) and 1 h of reperfusion to demonstrate that neutrophils occlude approximately 8% of the cerebral microvessels in the hypoperfused regions, resulting in the so-called no reflow phenomenon after removal of the original obstruction/lysis (del Zoppo et al. 1991). Thus, the contribution of neutrophils to brain damage may be underestimated in histopathologic studies of perfusion-fixed tissue, since a large component of leukocyte involvement is intravascular. After transient MCAO in rats, neutrophils can be detected in microvessels of the ischemic hemisphere as early as 30 min after the occlusion with a peak at 12 h, whereas intraparenchymal neutrophils were most numerous at 24 h (Garcia et al. 1994). The concurrence of parenchymal neutrophil accumulation and the presence of eosinophilic neurons suggests that neutrophils contribute to neuronal damage (Garcia et al. 1994; Zhang et al. 1994). Blood-borne monocytes and macrophages arrive late at the ischemic zone, but persist for a long time after the completion of the infarct. In primates, large numbers of monocytes/macrophages can be detected at the boundary of the infarct at days 7–16 (Garcia and Kamijyo 1974). Peak monocyte and macrophage accumulation in the CSF occurs between 3 and 7 days after stroke in humans (Sörnäs et al. 1972). Transient MCAO in rats also leads to the delayed appearance of monocytes/macrophages (Garcia et al. 1994). In addition to parenchymal infiltration, perivascular monocyte accumulation may propagate thrombosis in stroke (Holm and Hansson 1990). Lymphocytes can be found accumulating in the border zone of the infarct by 3 days after photothrombosis in rats (Jander et al. 1995).

Epidemiologic observations suggest a correlation between the white blood cell (WBC) count and the risk of stroke. In the Hiroshima/Nagasaki survivors' surveillance, WBC counts of greater than 10,000 per microliter were associated with a 1.6- to 2.6-fold greater risk of thrombotic stroke (Prentice et al. 1982). Importantly, the examination of differential WBC counts showed a statistically significant predictive power only for neutrophils. In clinical studies, stroke patients had a significantly higher mortality when their WBC counts were elevated on admission to the hospital (Lowe et al. 1983). High peripheral leukocyte counts at 3 days after stroke were associated with larger infarct size and impaired level of consciousness (Pozzilli et al. 1985a). After transient ischemic attacks, the risk of subsequent stroke is increased in the presence of elevated WBC counts (Harrison and Marshall 1987). In support of this notion, beneficial effects of neutropenia were observed in several experimental stroke models. Dutka et al. were first to show that mechlorethamine-induced neutropenia in dogs increased the requirement for air administration during air embolism-induced cerebral ischemia, and led to better blood flow and functional recovery after reperfusion (Dutka et al. 1989). In rabbits, mechlorethamine-induced neutrophil depletion completely eliminated post-ischemic hypoperfusion (Helps and Gorman 1991). Rabbits pretreated with antineutrophil antiserum had a better recovery of cerebral blood flow and a significantly smaller infarct size after autologous blood clot injection into the internal carotid artery (Bednar et al. 1991). Significant reductions in infarct volume were also noted in rats and mice depleted of neutrophils and subjected to transient focal cerebral ischemia (Matsuo et al. 1994; Connolly et al. 1996). On the other hand, some investigators have found no more than a bystander's role for neutrophils during focal cerebral ischemia and reperfusion (Hayward 1996). In global cerebral ischemia, a cyclophosphamide-induced 85% reduction in WBC count did not decrease the severity of reflow failure during reperfusion (Aspey et al. 1989). However, Grøgaard et al. found that a 95% reduction in peripheral blood neutrophils (and a 50% reduction in monocytes) by antibody pretreatment significantly attenuated the post-ischemic hypoperfusion after global ischemia (Grøgaard et al. 1989).

Antagonizing leukocyte adhesion molecules is another way of interfering with peripheral blood cell involvement in secondary brain damage after stroke. Treatment of baboons with an anti-CD18 monoclonal antibody was able to reduce leukocyte adhesion and improve reflow in microvessels after MCAO (Mori et al. 1992). A significant reduction in leukocyte accumulation and an improvement of neurological dysfunction was also observed in a rabbit model of spinal cord ischemia after anti-CD18 antibody treatment (Clark et al. 1991). Interestingly, benefit was shown even when the treatment was initiated as late as 30 min after reperfusion (Lindsberg et al. 1992). Intravenous application of an anti-CD11b monoclonal antibody in rats subjected to MCAO resulted in a dose-dependent reduction of infarct volume and parenchymal neutrophil infiltration, as well as in improved neurological function (Chen et al. 1994). The beneficial effects were observed even when the antibody was applied 60 min after reperfusion. Reduction of neutrophil infiltration and lesion size after reperfusion was also observed with anti-intercellular adhesion molecule (ICAM)-1 antibody treatment (Zhang et al. 1995a). Along these lines, null ICAM-1 mice displayed a more than threefold reduction in infarct volume compared with controls (Connolly et al. 1996). Treatment with antibodies against ICAM-1 and CD-18 also protected against brain damage induced by injection of autologous clots into the common carotid artery of rabbits (Bowes et al. 1995). It is noteworthy that anti-ICAM-1 antibody treatment 15 min after embolization prolonged the therapeutic window for tissue plasminogen activator (tPA) administration, suggesting that antiadhesion strategies might be used successfully in combination with thrombolytic therapy. Besides antibody treatment, synthetic fibronectin- and laminin-derived peptides, as well as heparin, also reduced leukocyte accumulation and led to improved outcome after transient ischemia/reperfusion in rats (Yanaka et al. 1996a,b, 1997).

8.3 Molecular and Cellular Mediators of Post-Ischemic Inflammatory Responses

The earliest observed changes in the ischemic brain include disruption of ionic gradients across membranes of excitable (neuronal) and non-excitable (glial) cells due to progressive energy failure. Release of potassium into the extracellular space triggers glutamate release, excitotoxic injury, and peri-infarct depolarizations (Dirnagl et al. 1999). Rapid activation of glial cells occurs prior to any obvious signs of neuronal death. Microglia, the resident macrophages of the central nervous system, are activated within 20 min of reperfusion following forebrain ischemia (Morioka al. 1991). They are believed to contribute to neuronal loss by releasing cytotoxic factors (Guilian and Vaca 1993), but their role in ischemia is controversial and some studies have suggested a neuroprotective function (Bruce et al. 1996). Astrocytes are also activated following ischemia, and express the β-amyloid precursor protein (Banati et al. 1995).

At the blood–endothelial interface, the leukocyte adhesion receptor, P-selectin, stored in alpha and dense granules of endothelial cells is translocated to the cell surface within 90 min after MCAO (Okada et al. 1994). P-selectin mediates the low-affinity "rolling" of neutrophils on the endothelium via binding of the lectin and epidermal growth factor domains to neutrophil surface carbohydrate ligands (Lasky 1992). E-selectins and L-selectins are expressed on endothelia and leukocytes, respectively. Importantly, expression of E-selectin is induced on intact microvessels after MCAO (Haring et al. 1996). Within minutes to hours after ischemia, the microvascular basal lamina, which is derived from extracellular matrix and functions as a barrier to cellular transit into the brain, is degraded by proteases. These events coincide with earliest evidence of neuron injury. Interestingly, tPA has recently been shown to potentiate *N*-methyl-d-aspartate-receptor signaling and excitotoxicity (Nicole et al. 2001). Matrix metalloproteinase (MMP)-2 is expressed in the primate striatum within 1 h following MCAO, and MMP-9 is most abundant in areas of hemorrhagic transformation (Heo et al. 1999). The receptors for laminins, collagen IV and fibronectin, the vascular matrix integrins, are also rapidly downregulated in response to ischemia. Expression of the integrin heterodimers, $\alpha_1\beta_1$ and $\alpha_3\beta_1$ on endothelia and $\alpha_6\beta_4$ on astrocytic end-feet, has decreased by 2 h post-MCAO (del Zoppo et al. 1996; Wagner et al. 1997). In contrast, the integrin $\alpha_v\beta_3$ is significantly upregulated in smooth muscle cells of arterioles, and may be involved in vascular remodeling after ischemia in association with vascular endothelial growth factor (Abumiya et al. 1999).

As the leukocyte encounters activated endothelia of postcapillary venules, firm adhesion is mediated by the interaction of members of the immunoglobulin superfamily, notably ICAM-1, ICAM-2, and vascular adhesion molecule (VCAM)-1, with leukocyte integrins. ICAM-1 is constitutively expressed at low levels by endothelial cells, but is substantially increased following ischemia. In a rat model of MCAO, ICAM-1 messenger RNA (mRNA) peaked at 10 h of reperfusion, and increased ICAM-1 surface protein expression was detectable in the ischemic hemisphere from 2 h to several days of reperfusion (Wang et al. 1994; Zhang et al. 1995b). After stroke in baboons, ICAM-1 was upregulated on the endothelium of postcapillary microvessels (Okada et al. 1994). Infarcted brain tissue from human stroke patients also harbored significantly more ICAM-1-positive microvessels compared with control subjects, while serum levels of ICAM-1 were unchanged or reduced (Sobel et al. 1990; Clark et al. 1993; Fassbender et al. 1995; Lindsberg et al. 1996). The relevance of ICAM-1 expression for leukocyte adhesion and secondary brain damage after stroke is underscored by the significant reduction in infarct volume observed in ICAM-1-null mice subjected to MCAO (Connolly et al. 1996). In contrast to ICAM-1, ICAM-2 is expressed at high levels on non-activated endothelial cells, and may thus play a role during rapidly occurring inflammation (Hallenbeck 1996). Induction of VCAM-1 has been shown in the ischemic hemisphere of stroke patients (Blann et al. 1999). Another member of the immunoglobulin superfamily, namely CD31 or platelet-endothelial cell adhesion molecule (PECAM)-1, appears to be crucial for transendothelial migration of leukocytes into the brain parenchyma (Hallenbeck 1996). While CD31 shows homophilic interaction, ICAM-1, ICAM-2, and VCAM-1 react with their complementary leukocyte integrin ligands, CD11b/CD18 which is the membranolytic attack complex (Mac)-1, CD11a/CD18 which is the lymphocyte function-related antigen (LFA)-1, and the very-late-after-activation antigen (VLA)-4, respectively. Upon neutrophil activation, CD11b/CD18 is mobilized from intracellular granules to the cell surface. CD18 adhesion molecules were found to be increased on circulating leukocytes within 12 h of symptom onset in human stroke patients (Fiszer et al. 1998). Elevated CD11a-expression was detected within 72 h after stroke (Kim et al. 1995). The significant contribution of leukocyte adhesion molecules to post-ischemic inflammatory injury is indicated by the results of recent experiments using anti-adhesion antibodies (see above). Moreover, mice deficient in Mac-1 or CD18, respectively showed a 26% and 53% reduction in infarct volume following transient MCAO compared to wildtype mice (Prestigiacomo et al. 1999; Soriano et al. 1999).

During inflammation, endothelial cells undergo functional changes that may lead to a prothrombotic state prepared by the "local Shwartzman reaction" (Hallenbeck et al. 1988). Expression of the procoagulant tissue factor, which is not normally present in endothelial cells, is induced by cytokines (Bevilacqua et al. 1984). Inhibition of protein S binding destabilizes the anticoagulant thrombomodulin-protein C–protein S system (Nawroth et al. 1986a). Fibrinolysis is inhibited by decreased tPA- and increased plasminogen activator inhibitor (PAI)-1-production (Bevilacqua et al. 1986). Thrombin-activated platelets bind firmly to fibrin via the integrin $\alpha_{IIb}\beta_3$, and platelet aggregation is promoted by thrombin and potentiated by adenosine diphosphate (ADP) interacting with the platelet P_{2Y1} purinoceptor (del Zoppo 1998). Finally, endothelial cells release vasoconstrictors, such as superoxide, thromboxane A_2 and endothelin-1. Endothelin-1 is so potent that it can reduce blood flow to ischemic levels (Macrae et al. 1993).

Among the signals that mediate the inflammatory response and the entry of leukocytes into the ischemic brain areas, cytokines and chemokines are considered to play the most prominent role. Cytokines are low-molecular-weight glycoproteins produced by most cells in the body in response to activation. They promote cell–cell interaction by binding to specific target-cell receptors, whose expression can be modulated by the cytokines themselves. Chemokines are small chemoattractant cytokines that contain characteristic N-terminal cysteine motifs, and act primarily in cellular trafficking. In the setting of cerebral ischemia and reperfusion, the two key cytokines are interleukin (IL)-1 and tumor necrosis factor (TNF)-α. IL-1 exists in two forms, IL-1α and IL-1β, which are generated from two separate genes. After transient MCAO and forebrain ischemia in rats, a rapid induction of IL-1β mRNA was observed within 15 min, reaching peak levels after 3–4 h and declining by 4 days (Minami et al. 1992; Buttini et al. 1994). Expression of the IL-1β protein was detected in macrophages and microglia, and increased steadily in the ischemic hemisphere during up to 48 h of MCAO in rats (Davies et al. 1999). In mice, IL-1β expression was found in endothelial cells and in microglia after MCAO (Zhanet al. 1998). A significant increase in IL-1β could also be detected in the CSF of stroke patients 2 days after the insult (Tarkowski et al. 1995). The increase in IL-1β production after focal ischemia in rats is accompanied by an early upregulation of IL-1-receptor type II mRNA and a delayed expression of IL-1-receptor type I mRNA (Wang et al. 1997). IL-1-receptor type II binds IL-1β with higher affinity than IL-1α. The biological effects of IL-1 are diverse, and include the induction of leukocyte adhesion molecules, such as ICAM-1 (Bevilacqua et al. 1985; Wang et al. 1995a), as well as the expression of prothrombotic factors, such as PAI-1 and tissue factor (Bevilacqua et al. 1984, 1986), on endothelium. IL-1β has also been shown to increase the production of endothelin-1 (Katabami et al. 1992). Applied intrathecally, IL-1β may act as a pyrogen and induce hyperthermia (Loddick and Rothwell 1996). Moreover, IL-1 is a potent enhancer of inflammation by stimulating the synthesis of other proinflammatory cytokines, such as IL-6, TNF-α, interferon (IFN)-γ, as well as its own synthesis (Touzani et al. 1999). The actions of IL-1 are tightly controlled in vivo, and the transcript of a naturally occurring antagonist, interleukin-1 receptor antagonist (IL-1ra), was found to be increased in the ischemic cortex from 6 h to several days after MCAO in rats (Wang et al. 1997). Several lines of evidence implicate IL-1 in brain damage after ischemia. First, intraventricular injection of IL-1β in rats subjected to transient or permanent MCAO exacerbated the infarct (Yamasaki et al. 1995a; Loddick and Rothwell 1996; Stroemer and Rothwell 1998). Second, administration of recombinant IL-1ra protected from damage even when applied 30 min after permanent MCAO in rats (Katabami et al. 1992; Relton et al. 1996). Third, intraventricular administration of anti-IL-1β neutralizing antibodies significantly reduced infarct areas after transient MCAO in rats (Yamasaki et al. 1995a). Finally, mice expressing a dominant negative mutant of IL-1β converting enzyme demonstrated smaller infarct sizes after permanent MCAO compared with controls (Friedlander et al. 1997).

TNF-α is another proinflammatory cytokine that has been implicated in brain injury. TNF-α and its receptors, p55 and p75, are expressed in the central nervous system (Smith and Baglioni 1992). Elevated TNF-α levels were detected in rats after excitoxic and traumatic head injury (Taupin et al. 1993; de Bock et al. 1996). Serum TNF-α levels were also increased in humans after severe head injury (Goodman et al. 1990). In rats, permanent MCAO induced TNF-α mRNA as early as 1 h after the onset of ischemia, with peak levels reached at 6–12 h (Liu et al. 1994; Zhai et al. 1997). TNF-α protein was first detected in microglia and neurons 30 min to 6 h after ischemia, and was later expressed by macrophages (Liu et al. 1994; Buttini et al. 1996). Peripheral blood cells isolated from human stroke patients showed increased TNF-α release for months after the stroke (Ferrarese et al. 1999). TNF-α exerts multiple biological actions, including conversion of endothelium to a prothrombotic surface (Nawroth et al. 1986b), induction of leukocyte adhesion to endothelial cells (Liu et al. 1994; Kim et al. 1992), increase in blood–brain barrier permeability (Wong et al. 1992) and macrophage activation (Riches et al. 1996). Soluble TNF receptors control the effects of TNF-α in vivo, and plasma levels of soluble TNF receptor protein-1 p55 (sTNF-RI) were upregulated in patients with acute stroke (Elneihoum et al. 1996). A role for TNF-α in ischemic brain damage comes from studies demonstrating that intraventricular administration of the cytokine prior to permanent or transient MCAO in spontaneously hypertensive rats significantly increased infarct size, edema and neurological deficit (Barone et al. 1997). The exacerbation could be reversed by pre-injection of anti-TNF-α antibody into the contralateral ventricle. Moreover, inhibition of increased endogenous TNF-α production in stroke-prone spontaneously hypertensive rats (Siren et al. 1992) by administration of monoclonal anti-TNF-α antibody or sTNF-RI reduced brain infarction even when applied several hours after MCAO (Dawson et al. 1996; Barone et al. 1997). In contrast, mice genetically deficient in TNF receptors unexpectedly had larger infarcts after transient ischemia and reperfusion (Bruce et al. 1996), indicating a pleiotropic action of TNF-α in cerebral ischemia.

Interleukin-6 is a hematopoietic cytokine, which is induced in animal models of transient ischemia and clinical stroke. IL-6 increased early after focal stroke in rats and peaked at 24 h (Legos et al. 2000). Expression was detected in neurons and microglia, but not in astrocytes (Suzuki et al. 1999). Although intraventricular injection of recombinant IL-6 significantly reduced ischemic brain damage after permanent MCAO in rats (Loddick et al. 1998), recent experimental evidence using IL-6-deficient mice suggests that IL-6 does not have a direct influence on acute ischemic injury (Clark et al. 2000). In clinical studies, elevated plasma levels of IL-6 were consistently associated with poor outcome after stroke (Fassbender et al. 1994; Beamer et al. 1995; Vila et al. 2000).

Recently, a number of experimental studies have addressed the contribution of inducible nitric oxide synthase (iNOS) and cyclooxygenase-2 (COX-2) to post-ischemic brain damage. These pro-inflammatory mediators are not expressed (or only at very low levels) in the normal brain, but were induced within 12–48 h after focal cerebral ischemia in rats (Iadecola et al. 1995; Iadecola et al. 1996; Miettinen et al. 1997; Nogawa et al. 1997). Expression of iNOS occurred in infiltrating inflammatory cells and cerebral blood vessels, while COX-2 was found in neurons and endothelia at the border of the ischemic territory. Induction of iNOS and COX-2 was also detected in the human brain after stroke (Forster et al. 1999; Iadecola et al. 1999). The detrimental effects of iNOS expression after cerebral ischemia are underscored by a 25%–30% reduction in infarct volume in iNOS-null mice and in rats treated with the relatively selective iNOS inhibitor, aminoguanidine, after MCAO (Zhang et al. 1996; Iadecola et al. 1997). Similarly, the COX-2 inhibitor, NS-398, reduced infarct volume by 30% after MCAO in rats (Nogawa et al. 1997). Given the spatial and temporal proximity of iNOS and COX-2 expression, nitric oxide produced by iNOS may activate COX-2, leading to the generation of toxic peroxynitrite, reactive oxygen species and prostanoids (Nogawa et al. 1998).

Among the cytokines with homeostatic action are transforming growth factor (TGF)-β, IL-4, and IL-10. TGF-β isoforms suppress translation of cytokine mRNAs, while IL-10 promotes their degradation (Bogdan et al. 1992). In rats subjected to permanent MCAO, expression of TGF-β1 mRNA was observed at 2 days and plateaued for up to 15 days after ischemia (Wang et al. 1995b). An early neuronal induction of TGF-β1 mRNA was followed by delayed astroglial TGF-β transcription in rat hippocampus after transient forebrain ischemia (Knuckey et al. 1996). Increased TGF-β1 mRNA expression and active TGF-β1 protein were also detected in ischemic strokes in humans, particularly in the penumbra (Krupinski et al. 1996). There is evidence to suggest that TGF-β1 protects mice from ischemic injury induced by permanent MCAO (Prehn et al. 1993). TGF-β is known to suppress endothelial adhesiveness for neutrophils (Gamble and Vadas 1988) and to deactivate macrophages (Tsunawaki et al. 1988). IL-10 mRNA peaked late at 2 days in the ischemic hemispheres of rats subjected to permanent MCAO, while IL4 mRNA was rapidly induced within 6 h (Li et al. 2001). In patients with ischemic stroke, IL-10 reached highest levels on day 3 in the CSF (Tarkowski et al. 1997), and strongly elevated numbers of IL-10-secreting (myelin basic protein-reactive) mononuclear cells were detected in peripheral blood (Pelidou et al. 1999). IL-4 was increased in the serum of stroke patients during the acute stage of disease (Kim et al. 2000). The anti-inflammatory nature of IL-10 is underscored by the fact that systemic IL-10 administration decreased infarct size in rats subjected to permanent MCAO (Spera et al. 1998). Moreover, IL-10 knockout mice had significantly larger brain infarcts following MCAO compared to controls (Grilli et al. 2000).

While cytokines play a mayor role in mediating adhesion of leukocytes to endothelium, they also stimulate chemokine production, which is required to attract blood-borne cells toward the ischemic area. Cytokine-induced neutrophil chemoattractant (CINC), a member of the IL-8 family, was shown to be maximally induced in the brain at 3–6 h after transient MCAO in rats (Yamasaki et al. 1995b; Yamagamet al. 1999), and at 12 h after permanent MCAO in spontaneously hypertensive rats (Liu et al. 1993). A peak serum concentration of CINC was observed 3 h after reperfusion in transient MCAO (Yamasaki et al. 1995b). CINC contains a structural motif that attracts predominantly neutrophils. The switch from neutrophil- to macrophage-chemoattraction may be mediated by macrophage inflammatory protein (MIP)-1, since this cytokine first attracted neutrophils and later macrophages in experimental meningitis (Saukkonen et al. 1990). MIP-1α and MIP-1β mRNA levels reached a peak of expression in microglia 8–16 h after MCAO (Gourmala et al. 1999). Monocyte chemoattractant protein (MCP)-1 followed the expression of CINC, and showed highest expression in astrocytes between 12–48 h after MCAO in rats (Gourmala et al. 1997; Yamagamet al. 1999). Finally, IFN-inducible protein-10 was found to be released by neurons and astrocytes after MCAO, and acts as a chemoattractant for macrophages and activated T lymphocytes (Wang et al. 1998a).

8.4 Potential Benefits of Inflammation

The above-mentioned acute detrimental effects of inflammation, and the transition from ischemic to inflammatory injury after stroke should not conceal the fact that inflammation is also intrinsically linked with repair and plasticity. In fact, implantation of macrophages that have been ex vivo exposed to nervous system antigens into transected rat spinal cord resulted in tissue repair and improved functional recovery of paraplegic animals (Rapalino et al. 1998). Moreover, regeneration of propriospinal neurons after spinal cord transection and human Schwann cell grafting in nude rats was most pronounced at sites of marked inflammation (Guest et al. 1997). Activated macrophages and microglia may enhance plasticity by their release of trophic factors, such as brain-derived neurotrophic factor and glial cell line-derived neurotrophic factor (Batchelor et al. 1999). T cell-mediated immunologic tolerance to myelin basic protein decreases stroke size through synthesis of TGF-β1 (Becker et al. 1997). Even the proinflammatory cytokine, IL-1β, may contribute to the limitation of secondary brain damage after stroke by inhibition of calcium currents, enhancement of γ-aminobutyric acid (GABA) activity and stimulation of nerve growth factor synthesis (Touzani et al. 1999). Both IL-1β and TNF-α are neuroprotective in the context of ischemic "pre-conditioning," i.e., the induction of ischemic tolerance by repeated administration of a sublethal stimulus (Ohtsuki et al. 1996; Nawashiro et al. 1997). IL-1β and TNF-α also upregulate the delayed expression of osteopontin in macrophages, which interacts with the integrin $\alpha_v\beta_3$ receptor on astrocytes, and is crucial for debridement and wound healing (Miyazaki et al. 1995; Wang et al. 1998b).

8.5 Therapeutic Perspective and Conclusion

Given the need for effective stroke treatment, and the encouraging results obtained with anti-leukocyte interventions in animal models of stroke, a human trial using an anti-ICAM-1 murine monoclonal antibody was initiated (ENLIMOMAB). In this placebo-controlled double-blind study, patients with acute ischemic stroke were randomized to receive either 160 mg+2–5×40 mg/d anti-ICAM-1 or placebo (The Enlimomab Acute Stroke Trial Investigators and Sherman 1997). Disappointingly, the trial resulted in more serious adverse events and in less improvement at 90 days in the Enlimomab group compared to control. While much can be said about the difficulties of immunoglobulin therapy (foreign protein challenge, complement activation, cell-surface cross-linking of antibodies resulting in oxidative burst, loss of natural negative feed-back), the failure of the trial also reminds of the complexities of the molecular and cellular interactions during post-ischemic inflammatory responses.

In summary, there is overwhelming evidence that inflammation plays a role in promoting damage after stroke. Reperfusion, which may occur spontaneously or when induced therapeutically, seems to accentuate the role of post-ischemic inflammatory reactions. Given the practical obstacles in acute stroke therapy, inflammation appears to be a promising therapeutic target with an extended therapeutic time window. Many of the existing anti-inflammatory compounds have favorable safety profiles when compared to stroke therapeutics which interfere with neurotransmission, such as glutamate receptor antagonists. In addition, clinical experience already exists with the use anti-inflammatory drugs for other disease conditions. However, it is also obvious that the role of inflammation in stroke is Janus-faced, and that besides deleterious effects (Del Zoppo et al. 2001), inflammation can also limit damage and be instrumental in tissue repair and remodeling (Feuerstein and Wang 2001). Future research, therefore, has to address the issue of whether and how to selectively target the harmful and enhance the beneficial effects of post-ischemic inflammatory reactions before embarking on further clinical trials.

Acknowledgements. Supported by the Hermann and Lilly Schilling Foundation and the Deutsche Forschungsgemeinschaft (DFG SFB 507-A5).

References

Abumiya T, Lucero J, Heo JH, Tagaya M, Koziol JA, Copeland BR, del Zoppo GJ (1999) Activated microvessels express vascular endothelial growth factor and integrin $\alpha_v\beta_3$ during focal cerebral ischemia. J Cereb Blood Flow Metab 19:1038–1050

Aspey BS, Jessimer C, Pereira S, Harrison MJ (1989) Do leukocytes have a role in the cerebral no-reflow phenomenon? J Neurol Neurosurg Psychiatr 52:526–528

Banati RB, Gehrmann J, Wiessner C, Hossmann KA, Kreutzberg GW (1995) Glial expression of the β-amyloid precursor protein (APP) in global ischemia. J Cereb Blood Flow Metab 15:647–654

Barone FC, Hillegass LM, Feuerstein GZ, Sarau HM, Clark RK, Griswold DE (1991) Myeloperoxidase activity assay and histology verify polymorphonuclear leukocytes (PMN) in focal ischemia: effect of reperfusion. Stroke 22:131

Barone FC, Arvin B, White RF, Miller A, Webb CL, Willette RN, Lysko PG, Feuerstein GZ (1997) Tumor necrosis factor-alpha. A mediator of focal ischemic brain injury. Stroke 28:1233–1244

Batchelor PE, Liberatore GT, Wong JY, Porritt MJ, Frerichs F, Donnan GA, Howells DW (1999) Activated macrophages and microglia induce dopaminergic sprouting in the injured striatum and express brain-derived neurotrophic factor and glial cell line-derived neurotrophic factor. J Neurosci 19:1708–1716

Beamer NB, Coull BM, Clark WM, Hazel JS, Silberger JR (1995) Interleukin-6 and interleukin-1 receptor antagonist in acute stroke. Ann Neurol 37:800–805

Becker KJ, McCarron RM, Ruetzler C, Laban O, Sternberg E, Flanders KC, Hallenbeck JM (1997) Immunologic tolerance to myelin basic protein decreases stroke size after transient focal cerebral ischemia. Proc Natl Acad Sci USA 94:10873–10878

Bednar MM, Raymond S, McAuliffe T, Lodge PA, Gross CE (1991) The role of neutrophils and platelets in a rabbit model of thrombembolic stroke. Stroke 22:44–50

Bevilacqua MP, Pober JS, Majeau GR, Cotran RS, Gimbrone MA Jr (1984) Interleukin-1 (IL-1) induces biosynthesis and cell surface expression of procoagulant activity in human vascular endothelial cells. J Exp Med 160:618–623

Bevilacqua MP, Pober JS, Wheeler ME, Cotran RS, Gimbrone MA Jr (1985) Interleukin-1 acts on cultured human vascular endothelium to increase the adhesion of polymorphonuclear leukocytes, monocytes and related leukocyte cell lines. J Clin Invest 76:2003–2011

Bevilacqua MP, Schleef RR, Gimbrone MA Jr, Loskutoff DJ (1986) Regulation of the fibrinolytic system of cultured human vascular endothelium by interleukin 1. J Clin Invest 78:587–591

Blann A, Kumar P, Krupinski J, McCollum C, Beevers DG, Lip GY (1999) Soluble intercelluar adhesion molecule-1, E-selectin, vascular cell adhesion molecule-1 and von Willebrand factor in stroke. Blood Coagul Fibrinolysis 10:277–284

Bogdan C, Paik J, Vodovotz Y, Nathan C (1992) Contrasting mechanisms for suppression of macrophage cytokine release by transforming growth factor-β and interleukin-10. J Biol Chem 267:23301–23308

Bowes MP, Rothlein R, Fagan SC, Zivin JA (1995) Monoclonal antibodies preventing leukocyte activation reduce experimental neurologic injury and enhance efficacy of thrombolytic therapy. Neurology 45:815–819

Bruce A, Boling W, Kindy MS, Peschon J, Kraemer PJ, Carpenter MK, Holtsberg FW, Mattson MP (1996) Altered neuronal and microglial responses to excitotoxic and ischemic brain injury in mice lacking TNF receptors. Nat Medicine 2:788–794

Buttini M, Sauter A, Boddeke HW (1994) Induction of interleukin-1 beta mRNA after focal cerebral ischemia in the rat. Mol Brain Res 23:126–134

Buttini M, Appel K, Sauter A, Gebicke-Haerter PJ, Boddeke HW (1996) Expression of tumor necrosis factor alpha after focal cerebral ischemia in the rat. Neuroscience 71:1–16

Chen H, Chopp M, Zhang RL, Bodzin G, Chen Q, Rusche JR, Todd RF 3rd (1994) Anti-CD11b monoclonal antibody reduces ischemic cell damage after transient focal cerebral ischemia in rat. Ann Neurol 35:447–452

Chuaqui R, Tapia J (1993) Histologic assessment of the age of recent brain infarcts in man. J Neuropathol Exp Neurol 52:481–489

Clark WM, Madden KP, Lyden PD, Zivin JA (1991) Reduction of central nervous system ischemic injury in rabbits using leukocyte adhesion antibody treatment. Stroke 22:877–883

Clark WM, Coull BM, Briley DP, Mainolfi E, Rothlein R (1993) Circulating intercellular adhesion molecule-1 levels and neutrophil adhesion in stroke. J Neuroimmunol 44:123–125

Clark WM, Rinker LG, Lessov NS, Hazel K, Hill JK, Stenzel-Poore M, Eckenstein F (2000) Lack of interleukin-6 expression is not protective against focal central nervous system ischemia. Stroke 31:1715–1720

Connolly ES Jr, Winfree CJ, Springer TA, Naka Y, Liao H, Yan SD, Stern DM, Solomon RA, Gutierrez-Ramos JC, Pinsky DJ (1996) Cerebral protection in homozygous null ICAM-1 mice after middle cerebral artery occlusion: role of neutrophil adhesion in the pathogenesis of stroke. J Clin Invest 97:209–216

Davies CA, Loddick SA, Toulmond S, Stroemer RP, Hunt J, Rothwell NJ (1999) The progression and topographic distribution of interleukin-1beta expression after permanent middle cerebral artery occlusion in the rat. J Cereb Blood Flow Metab 19:87–98

Dawson DA, Martin D, Hallenbeck JM (1996) Inhibition of tumor necrosis factor-alpha reduces focal cerebral ischemic injury in the spontaneously hypertensive rat. Neurosci Lett 218:41–44

de Bock F, Dornand J, Rondouin G (1996) Release of TNF alpha in the rat hippocampus following epileptic seizures and excitotoxic neuronal damage. Neuroreport 7:1125–1129

del Zoppo GJ (1998) The role of platelets in ischemic stroke. Neurology 51:S9-S14

del Zoppo GJ, Pessin MS, Mori E, Hacke W (1991) Polymorphonuclear leukocytes occlude capillaries following middle cerebral artery occlusion and reperfusion in baboons. Stroke 22:1276–1283 del Zoppo GJ, Haring H-P, Tagaya M, Wagner S, Akamine P, Hamann GF (1996) Loss of $\alpha_1\beta_1$ integrin immunoreactivity on cerebral microvessels and astrocytes following focal cerebral ischemia/reperfusion. Cerebrovasc Dis 6:9

del Zoppo GJ, Becker KJ, Hallenbeck JM (2001) Inflammation after stroke: is it harmful? Arch Neurol 58:669–672

Dirnagl U, Iadecola C, Moskowitz MA (1999) Pathobiology of ischaemic stroke: an integrated view. Trends Neurosci 22:391–397

Dutka AJ, Kochanek PM, Hallenbeck JM (1989) Influence of granulocytopenia on canine cerebral ischemia induced by air embolism. Stroke 20:390–395

Enlimomab Acute Stroke Trial Investigators The, Sherman DG (1997) The Enlimomab Acute Stroke Trial: final results. Neurology 48:A270

Elneihoum AM, Falke P, Axelsson L, Lundberg E, Lindgarde F, Ohlsson K (1996) Leukocyte activation detected by increased plasma levels of inflammatory mediators in patients with ischemic cerebrovascular diseases. Stroke 27:1734–1738

Fassbender K, Rossol S, Kammer T, Daffertshofer M, Wirth S, Dollman M, Hennerici M (1994) Proinflammatory cytokines in serum of patients with acute cerebral ischemia: kinetics of secretion and relation to the extent of brain damage and outcome of disease. J Neurol Sci 122:135–139

Fassbender K, Mossner R, Motsch L, Kischka U, Grau A, Hennerici M (1995) Circulating selectin- and immunoglobulin-type adhesion molecules in acute ischemic stroke. Stroke 26:1361–1364

Ferrarese C, Mascarucci P, Zoia C, Cavarretta R, Frigo M, Begni B, Sarinella F, Frattola L, De Simoni MG (1999) Increased cytokine release from peripheral blood cells after acute stroke. J Cereb Blood Flow Metab 19:1004–1009

Feuerstein GZ, Wang X (2001) Inflammation and stroke: benefits without harm? Arch Neurol 58:672–673

Fiszer U, Korczak-Kowalska G, Palasik W, Korlak J, Gorski A, Czlonkowska A (1998) Increased expression of adhesion molecule CD18 (LFA-1beta) on the leukocytes of peripheral blood in patients with acute ischemic stroke. Acta Neurol Scand 97:221–224

Forster C, Clark HB, Ross ME, Iadecola C (1999) Inducible nitric oxide synthase expression in human cerebral infarcts. Acta Neuropathol 97:215–220

Friedlander RM, Gagliardini V, Hara H, Fink KB, Li W, MacDonald G, Fishman MC, Greenberg AH, Moskowitz MA, Yuan J (1997) Expression of a dominant negative mutant of interleukin-1 beta converting enzyme in transgenic mice prevents neuronal cell death induced by trophic factor withdrawal and ischemic brain injury. J Exp Med 185:933–940

Gamble JR, Vadas MA (1988) Endothelial adhesiveness for blood neutrophils is inhibited by transforming growth factor-β. Science 242:97—99

Garcia JH, Kamijyo Y (1974) Cerebral infarction: evolution of histopathological changes after occlusion of a middle cerebral artery in primates. J Neuropathol Exp Neurol 33:408–421

Garcia JH , Liu KF, Yoshida Y, Lian J, Chen S, del Zoppo GJ (1994) Influx of leukocytes and platelets in an evolving brain infarct (Wistar rat). Am J Pathol 144:188–199

Goodman JC , Robertson CS, Grossman RG, Narayan RK (1990) Elevation of tumor necrosis factor in head injury. J Neuroimmunol 30:213-217

Gourmala NG , Buttini M, Limonta S, Sauter A, Boddeke HW (1997) Differential and time-dependent expression of monocyte chemoattractant protein-1 mRNA by astrocytes and macrophages in rat brain: effects of ischemia and peripheral lipopolysaccharide administration. J Neuroimmunol 74:35–44

Gourmala NG , Limonta S, Bochelen D, Sauter A, Boddeke HW (1999) Localization of macrophage inflammatory protein: macrophage inflammatory protein-1 expression in rat brain after peripheral administration of lipopolysaccharide and focal cerebral ischemia. Neuroscience 88:1255–1266

Grilli M , Barbieri I, Basudev H, Brusa R, Casati C, Lozza G, Ongini E (2000) Interleukin-10 modulates neuronal threshold of vulnerability to ischaemic damage. Eur J Neurosci 12:2265–2272

Grøgaard B , Schurer L, Gerdin B, Arfors KE (1989) Delayed hypoperfusion after incomplete forebrain ischemia in the rat: the role of polymorphonuclear leukocytes. J Cereb Blood Flow Metab 9:500–505

Guest JD , Rao A, Olson L, Bunge MB, Bunge RP (1997) The ability of human Schwann cell grafts to promote regeneration in the transected nude rat spinal cord. Exp Neurol 148:502–522

Guilian D, Vaca K (1993) Inflammatory glia mediate delayed neuronal damage after ischemia in the central nervous system. Stroke 24:84—90

Hallenbeck JM (1996) Inflammatory reactions at the blood-endothelial interface in acute stroke. Adv Neurol 71:281–97

Hallenbeck JM, Dutka AJ, Kochanek PM, Siren A, Pezeshkpour GH, Feuerstein G (1988) Stroke risk factors prepare rat brainstem tissues for modified local Schwartzman reaction. Stroke 19:863–869

Haring H-P, Berg EL, Tsurushita N, Tagaya M, del Zoppo GJ (1996) E-selectin appears in non-ischemic tissue during experimental focal cerebral ischemia. Stroke 27:1391–1392

Harrison M, Marshall J (1987) Does the peripheral blood leukocyte count predict the risk of transient ischemic attacks and strokes? J Neurol Neurosurg Psychiatry 50:1558–1559

Hayward NJ (1996) Lack of evidence for neutrophil participation during infarct formation following focal cerebral ischemia in the rat. Exp Neurol 139:188–202

Helps SC, Gorman DF (1991) Air embolism of the brain in rabbits pretreated with mechlorethamine. Stroke 22:351–354

Heo J, Lucero J, Abumiya T, Koziol JA, Copeland BR, del Zoppo GJ (1999) Matrix metalloproteinases increase very early during experimental focal ischemia. J Cereb Blood Flow Metab 19:624–633

Holm J, Hansson GK (1990) Cellular and immunologic features of carotid artery disease in man and experimental animal models. Eur J Vasc Surg 4:49–55

Iadecola C, Zhang F, Xu S, Casey R, Ross ME (1995) Inducible nitric oxide synthase gene expression in brain following cerebral ischemia. J Cereb Blood Flow Metab 15:378–384

Iadecola C, Zhang F, Casey R, Clark HB, Ross ME (1996) Inducible nitric oxide synthase gene expression in vascular cells after transient focal cerebral ischemia. Stroke 27:1373–1380

Iadecola C, Zhang F, Casey R, Nagayama M, Ross ME (1997) Delayed reduction of ischemic brain injury and neurological deficits in mice lacking the inducible nitric oxide synthase gene. J Neurosci 17:9157–9164

Iadecola C, Forster C, Nogawa S, Clark HB, Ross ME (1999) Cyclooxygenase-2 immunoreactivity in the human brain following cerebral ischemia. Acta Neuropathol 98:9–14

Jander S, Kraemer M, Schroeter M, Witte OW, Stoll G (1995) Lymphocytic infiltration and expression of intercellular adhesion molecule-1 in photochemically induced ischemia of the rat cortex. J Cereb Blood Flow Metab 15:42–51

Katabami T, Shimizu M, Okano K, Yano Y, Nemoto K, Ogura M, Tsukamoto T, Suzuki S, Ohira K, Yamada Y, et al (1992) Intracellular signal transduction for interleukin-1 beta-induced endothelin production in human umbilical vein endothelial cells. Biochem Biophys Res Commun 188:565–570

Kim HM, Shin HY, Jeong HJ, An HJ, Kim NS, Chae HJ, Kim HR, Song HJ, Kim KY, Baek SH, Cho KH, Moon BS, Lee YM (2000) Reduced IL-2 but elevated IL-4, IL-6, and IgE serum levels in patients with cerebral infarction during the acute stage. J Mol Neurosci 14:191–196

Kim JS, Chopp M, Chen H, Levine SR, Carey JL, Welch KM (1995) Adhesive glycoproteins CD11a and CD18 are upregulated in the leukocytes from patients with ischemic stroke and transient ischemic attacks. J Neurol Sci 128:45–50

Kim KS, Wass CA, Cross AS, Opal SM (1992) Modulation of blood-brain barrier permeability by tumor necrosis factor and antibody to tumor necrosis factor in the rat. Lymphokine Cytokine Res 11:293–298

Knuckey NW, Finch P, Palm DE, Primiano MJ, Johanson CE, Flanders KC, Thompson NL (1996) Differential neuronal and astrocytic expression of transforming growth factor beta isoforms in rat hippocampus following transient forebrain ischemia. Mol Brain Res 40:1–14

Kochanek PM, Dutka AJ, Hallenbeck JM (1987) Indomethacin, prostacyclin, and heparin improve postischemic cerebral blood flow without affecting early postischemic granulocyte accumulation. Stroke 18:634–637

Krupinski J, Kumar P, Kumar S, Kaluza J (1996) Increased expression of TGF-beta 1 in brain tissue after ischemic stroke in humans. Stroke 27:852–857

Lasky LA (1992) Selectins: interpreters of cell-specific carbohydrate information during inflammation. Science 258:964–969

Legos JJ, Whitmore RG, Erhardt JA, Parsons AA, Tuma RF, Barone FC (2000) Quantitative changes in interleukin proteins following focal stroke in the rat. Neurosci Lett 24:189–192

Li HL, Kostulas N, Huang YM, Xiao BG, van der Meide P, Kostulas V, Giedraitas V, Link H (2001) IL-17 and IFN-gamma mRNA expression is increased in the brain and systemically after permanent middle cerebral artery occlussion in the rat. J Neuroimmunol 116:5–14

Lindsberg PJ, Siren AL, Feuerstein GZ, Hallenbeck JM (1991) Post-ischemic antagonism of neutrophil adherence has an acute therapeutic effect on functional recovery in the deteriorating stroke model in rabbits. J Cereb Blood Flow Metab 11:S754

Lindsberg PJ, Carpen O, Paetau A, Karjalainen-Lindsberg ML, Kaste M (1996) Endothelial ICAM-1 expression associated with inflammatory cell response in human ischemic stroke. Circulation 94:939–945

Liu T, Young PR, McDonnell PC, White RF, Barone FC, Feuerstein GZ (1993) Cytokine-induced neutrophil chemoattractant mRNA expressed in cerebral ischemia. Neurosci Lett 164:125–128

Liu T, Clark RK, McDonnell PC, Young PR, White RF, Barone FC, Feuerstein GZ (1994) Tumor necrosis factor-alpha expression in ischemic neurons. Stroke 25:1481—1488

Loddick SA, Rothwell NJ (1996) Neuroprotective effects of human recombinant interleukin-1 receptor antagonist in focal cerebral ischaemia in the rat. J Cereb Blood Flow Metab 16:932–940

Loddick SA, Turnbull AV, Rothwell NJ (1998) Cerebral interleukin-6 is neuroprotective during permanent focal cerebral ischemia in the rat. J Cereb Blood Flow Metab 18:176–179

Lowe GDO, Jaap AJ, Forbes CD (1983) Relation of atrial fibrillation and high haematocrit to mortality in acute stroke. Lancet 1:784–786

Macrae IM, Jaap AJ, Forbes CD (1993) Endothelin-1-induced reductions in cerebral blood flow: dose dependency, time course, and neuropathological consequences. J Cereb Blood Flow Metab 13:276–284

Matsuo Y, Onodera H, Shiga Y, Nakamura M, Ninomiya M, Kihara T, Kogure K (1994) Correlation between myeloperoxidase-quantified neutrophil accumulation and ischemic brain injury in rat: effect of neutrophil depletion. Stroke 25:1469–1475

Miettinen S, Fusco FR, Yrjanheikki J, Keinanen R, Hirvonen T, Roivainen R, Narhi M, Hokfelt T, Koistinaho J (1997) Spreading depression and focal brain ischemia induce cyclooxygenase-2 in cortical neurons through N-methyl-D-aspartic acid-receptors and phospholipase A2. Proc Natl Acad Sci USA 94:6500–6505

Minami M, Kuraishi Y, Yabuuchi K, Yamazaki A, Satoh M (1992) Induction of interleukin-1 beta mRNA in rat brain after transient forebrain ischemia. J Neurochem 58:390–392

Miyazaki T, Tashiro T, Higuchi Y, Setoguchi M, Yamamoto S, Nagai H, Nasu M, Vassalli P (1995) Expression of osteopontin in a macrophage cell line and in transgenic mice with pulmonary fibrosis resulting from the lung expression of a tumor necrosis factor-alpha transgene. Ann NY Acad Sci 760:334–341

Mori E, del Zoppo GJ, Chambers JD, Copeland BR, Arfors KE (1992) Inhibition of polymorphonuclear leukocyte adherence suppresses no-reflow after focal cerebral ischemia in baboons. Stroke 23:712–718

Morioka, Kalehua AN, Streit WJ (1991) The microglial reaction in the dorsal hippocampus following transient forebrain ischemia. J Cereb Blood Flow Metab 10:966–973

The National Institute of Neurological Disorders and Stroke rt-PA Stroke Study Group (1995) Tissue plasminogen activator for acute ischemic stroke. N Engl J Med 333:1581–1587

Nawashiro H, Tasaki K, Ruetzler CA, Hallenbeck JM (1997) TNF-alpha pretreatment induces protective effects against focal cerebral ischemia in mice. J Cereb Blood Flow Metab 17:483–490

Nawroth PP, Handley DA, Esmon CT, Stern DM (1986a) Interleukin-1 induces endothelial cell procoagulant while suppressing cell surface anticoagulant activity. Proc Natl Acad Sci USA 83:3460–3464

Nawroth PP, Bank I, Handley D, Cassimeris J, Chess L, Stern D (1986b) Modulation of endothelial cell hemostatic properties by tumor necrosis factor. J Exp Med 163:740–745

Nicole O, Docagne F, Ali C, Margaill I, Carmeliet P, MacKenzie ET, Vivien D, Buisson A (2001) The proteolytic activity of tissue-plasminogen activator enhances NMDA receptor-mediated signaling. Nat Med 7:59–64

Nogawa S, Zhang F, Ross ME, Iadecola C (1997) Cyclo-oxygenase-2 gene expression in neurons contributes to ischemic brain damage. J Neurosci 17:2746–2755

Nogawa S, Forster C, Zhang F, Nagayama M, Ross ME, Iadecola C (1998) Interaction between inducible nitric oxide synthase and cyclooxygenase-2 after cerebral ischemia. Proc Natl Acad Sci USA 95:10966–10971

Ohtsuki T, Ruetzler CA, Tasaki K, Hallenbeck JM (1996) Interleukin-1 mediates induction of tolerance to global ischemia in gerbil hippocampal CA1 neurons. J Cereb Blood Flow Metab 16:1137–1142

Okada Y, Copeland BR, Mori E, Tung MM, Thomas WS, del Zoppo GJ (1994) P-selectin and intercellular adhesion molecule-1 expression after focal brain ischemia and reperfusion. Stroke 25:202–211

Pelidou SH, Kostulas N, Matusevicius D, Kivisakk P, Kostulas V, Link H (1999) High levels of IL-10 secreting cells are present in blood in cerebrovascular diseases. Eur J Neurol 6:437–442

Pozzilli C, Lenzi GL, Argentino C, Carolei A, Rasura M, Signore A, Bozzao L, Pozzilli P (1985a) Imaging of leukocytic infiltration in human cerebral infarcts. Stroke 16:251–255

Pozzilli C, Lenzi GL, Argentino C, Bozzao L, Rasura M, Giubilei F, Fieschi C (1985b) Peripheral white blood cell count in cerebral ischemic infarction. Acta Neurol Scand 71:396–400

Prehn JH, Backhauss C, Krieglstein J (1993) Transforming growth factor-beta 1 prevents glutamate neurotoxicity in rat neocortical cultures and protects mouse neocortex from ischemic injury in vivo. J Cereb Blood Flow Metab 13:521–525

Prentice RL, Szatrowski TP, Kato H, Mason MW (1982) Leukocyte counts and cerebrovascular disease. J Chronic Dis 35:703–714

Prestigiacomo CJ, Kim SC, Connolly ES Jr, Liao H, Yan SF, Pinsky DJ (1999) CD18-mediated neutrophil recruitment contributes to the pathogenesis of reperfused but not nonreperfused stroke. Stroke 30:1110–1117

Rapalino O, Lazarov-Spiegler O, Agranov E, Velan GJ, Yoles E, Fraidakis M, Solomon A, Gepstein R, Katz A, Belkin M, Hadani M, Schwartz M (1998) Implantation of stimulated homologous macrophages results in partial recovery of paraplegic rats. Nat Medicine 4:814–821

Relton JK, Martin D, Thompson RC, Russell DA (1996) Peripheral administration of interleukin-1 receptor antagonist inhibits brain damage after focal cerebral ischemia in the rat. Exp Neurol 138:206–213

Riches DW, Chan ED, Winston BW (1996) TNF-alpha-induced regulation and signalling in macrophages. Immunobiology 195:477–490

Saukkonen K, Sande S, Cioffe C, Wolpe S, Sherry B, Cerami A, Tuomanen E (1990) The role of cytokines in the generation of inflammation and tissue damage in experimental gram-positive meningitis. J Exp Med 171:439–448

Siren AL, Heldman E, Doron D, Lysko PG, Yue TL, Liu Y, Feuerstein G, Hallenbeck JM (1992) Release of proinflammatory and prothrombotic mediators in the brain and peripheral circulation in spontaneously hypertensive and normotensive Wistar-Kyoto rats. Stroke 23:1643–1650

Smith RA, Baglioni C (1992) Characterization of TNF receptors. Immunol Ser 56:131–147

Sobel RA, Mitchell ME, Fondren G (1990) Intercellular adhesion molecule-1 (ICAM-1) in cellular immune reactions in the human central nervous system. Am J Pathol 136:1309–1316

Soriano SG, Coxon A, Wang YF, Frosch MP, Lipton SA, Hickey PR, Mayadas TN (1999) Mice deficient in Mac-1 (CD11b/CD18) are less susceptible to cerebral ischemia/reperfusion injury. Stroke 30:134–139

Sörnäs R, Ostlund H, Muller R (1972) Cerebrospinal fluid cytology after stroke. Arch Neurol 26:489–501

Spera PA, Ellison JA, Feuerstein GZ, Barone FC (1998) IL-10 reduces rat brain injury following focal stroke. Neurosci Lett 251:189–192

Stroemer RP, Rothwell NJ (1998) Exacerbation of ischemic brain damage by localized striatal injection of interleukin-1 beta in the rat. J Cereb Blood Flow Metab 18:833–839

Suzuki S, Tanaka K, Nogawa S, Nagata E, Ito D, Dembo T, Fukuuchi Y (1999) Temporal profile and cellular localization of interleukin-6 protein after focal ischemia in rats. J Cereb Blood Flow Metab 19:1256–1262

Tarkowski E, Rosengren L, Blomstrand C, Wikkelso C, Jensen C, Ekholm S, Tarkowski A (1995) Early intrathecal production of interleukin-6 predicts the size of brain lesion in stroke. Stroke 26:1393–1398

Tarkowski E, Rosengren L, Blomstrand C, Wikkelso C, Jensen C, Ekholm S, Tarkowski A (1997) Intrathecal release of pro- and anti-inflammatory cytokines during stroke. Clin Exp Immunol 110:492–499

Taupin V, Toulmond S, Serrano A, Benavides J, Zavala F (1993) Increase in IL-6 and TNF levels in rat brain following traumatic lesion. Influence of pre- and post-traumatic treatment with Ro54866, a peripheral-type (p site) ligand. J Neuroimmunol 42:177–185

Touzani O, Boutin H, Chuquet J, Rothwell N (1999) Potential mechanisms of interleukin-1 involvement in cerebral ischemia. J Neuroimmunol 100:203–215

Tsunawaki S, Sporn M, Ding A, Nathan C (1988) Deactivation of macrophages by transforming growth factor-β. Nature 334:260–262

Vila N, Castillo J, Davalos A, Chamorro A (2000) Proinflammatory cytokines and early neurological worsening in ischemic stroke. Stroke 31:2325–2329

Wagner S, Tagaya M, Koziol JA, Quaranta V, del Zoppo GJ (1997) Rapid disruption of an astrocyte interaction with the extracellular matrix mediated by $\alpha_6\beta_4$ during focal cerebral ischemia/reperfusion. Stroke 28:858–865

Wang X, Siren AL, Liu Y, Yue TL, Barone FC, Feuerstein GZ (1994) Upregulation of intercellular adhesion molecule 1 (ICAM-1) on brain microvascular endothelial cells in rat ischemic cortex. Mol Brain Res 26:61–68

Wang X, Feuerstein GZ, Gu JL, Lysko PG, Yue TL (1995a) Interleukin-1 beta induces expression of adhesion molecules in human vascular smooth muscle cells and enhances adhesion of leukocytes to smooth muscle cells. Atherosclerosis 115:89–98

Wang X, Yue TL, White RF, Barone FC, Feuerstein GZ (1995b) Transforming growth factor-beta 1 exhibits delayed gene expression following cerebral ischemia. Brain Res Bull 36:607–609

Wang X, Barone FC, Aiyar NV, Feuerstein GZ (1997) Interleukin-1 receptor and receptor antagonist gene expression after focal stroke in rats. Stroke 28:155–161

Wang X, Ellison JA, Siren AL, Lysko PG, Yue TL, Barone FC, Shatzman A, Feuerstein GZ (1998a) Prolonged expression of interferon-inducible protein-10 in ischemic cortex after permanent occlusion of the middle cerebral artery in rat. J Neurochem 71:1194–1204

Wang X, Louden C, Yue TL, Ellison JA, Barone FC, Solleveld HA, Feuerstein GZ (1998b) Delayed expression of osteopontin after focal stroke in the rat. J Neurosci 18:2075–2083

Wong D, Dorovini-Zis K (1992) Up-regulation of intercellular adhesion molecule-1 (ICAM-1) expression in primary cultures of human brain microvessel endothelial cells by cytokines and lipopolysaccharide. J Neuroimmunol 39:11–21

Yamagami S, Tamura M, Hayashi M, Endo N, Tanabe H, Katsuura Y, Komoriya K (1999) Differential production of MCP-1 and cytokine-induced neutrophil chemoattractant in the ischemic brain after transient focal ischemia in rats. J Leukoc Biol 65:744–749

Yamasaki Y, Matsuura N, Shozuhara H, Onodera H, Itoyama Y, Kogure K (1995a) Interleukin-1 as a pathogenetic mediator of ischemic brain damage in rats. Stroke 26:676–680

Yamasaki Y, Matsuo Y, Matsuura N, Onodera H, Itoyama Y, Kogure K (1995b) Transient increase of cytokine-induced neutrophil chemoattractant, a member of the interleukin-8 family, in ischemic brain areas after focal ischemia in rats. Stroke 26:318–323

Yanaka K, Camarata PJ, Spellman SR, McCarthy JB, Furcht LT, Low WC, Heros RC (1996a) Neuronal protection from cerebral ischemia by synthetic fibronectin peptides to leukocyte adhesion molecules. J Cereb Blood Flow Metab 16:1120–1125

Yanaka K, Spellman SR, McCarthy JB, Low WC, Camarata PJ (1996b) Reduction of brain injury using heparin to inhibit leukocyte accumulation in a rat model of transient focal cerebral ischemia: I-protective mechanisms. J Neurosurg 85:1108–1112

Yanaka K, Camarata PJ, Spellman SR, Skubitz AP, Furcht LT, Low WC (1997) Laminin peptide ameliorates brain injury by inhibiting leukocyte accumulation in a rabbit model of transient focal cerebral ischemia J Cereb Blood Flow Metab 17:605–611

Zhai QH, Futrell N, Chen FJ (1997) Gene expression of IL-10 in relationship to TNF-alpha, IL-1beta and IL-2 in the rat brain following middle cerebral artery occlusion. J Neurol Sci 25:119–124

Zhang F, Casey RM, Ross ME, Iadecola C (1996) Aminoguanidine ameliorates and L-arginine worsens brain damage from intraluminal middle cerebral artery occlusion. Stroke 27:317–323

Zhang RL, Chopp M, Chen H, Garcia JH (1994) Temporal profile of ischemic tissue damage, neutrophil response, and vascular plugging following permanent and transient (2 h) middle cerebral artery occlusion in the rat. J Neurol Sci 125:3–10

Zhang RL, Chopp M, Jiang N, Tang WX, Prostak J, Manning AM, Anderson DC (1995a) Anti-intercellular adhesion molecule-1 antibody reduces ischemic cell damage after transient but not permanent middle cerebral artery occlusion in the Wistar rat. Stroke 26:1438–1443

Zhang RL, Chopp M, Zaloga C, Zhang ZG, Jiang N, Gautam SC, Tang WX, Tsang W, Anderson DC, Manning AM (1995b) The temporal profiles of ICAM-1 protein and mRNA expression after transient MCA occlusion in the rat. Brain Res 682:182–188

Zhang Z, Chopp M, Goussev A, Powers C (1998) Cerebral vessels express interleukin 1beta after focal cerebral ischemia. Brain Res 784:210–217

9 Neuroinflammation in Alzheimer's Disease: Potential Targets for Disease-Modifying Drugs

M. Hüll, H. Hampel

9.1 Neuropathology of Alzheimer's Disease 159
9.2 Components of Neuroinflammation in the Central Nervous System 160
9.3 Microglial Activation in Alzheimer's Disease 161
9.4 Cytokines, Acute-Phase Proteins and Complement Components . 162
9.5 Cyclooxygenase 1/2 and the Production of Prostaglandins 163
9.6 Neuroinflammation in Animal Models and Cell Cultures 164
9.7 Vaccination Studies in Animal Models 165
9.8 Epidemiological and Clinical Data on Anti-Inflammatory Drugs . 166
9.9 Neuroinflammation and Possible Drug Targets 167
9.10 Consideration for Further Research 170
References 171

9.1 Neuropathology of Alzheimer's Disease

Since the earliest description of Alzheimer's disease (AD) at the beginning of the last century, the intra- and extracellular deposition of proteins and the alterations in neuronal and glial morphology have been consistently detected in the brains of demented patients. During the following decades with the enormous developments in molecular biology and electron microscopy, the most obvious histopathological elements of AD have been described in more details on the molecular level and the topographical distribution. While intracellular neurofibrillary tangles and extracellular amyloid deposits are found in varying amounts in neuropathological samples of AD, numeric loss of synapses and neuronal cell death are most closely correlated to cognitive decline. However, despite the extensive characterisation of intracellular neurofibrillary tangles and extracellular amyloid deposits on the molecular level, the pathophysiological relationship between these four type of lesions – neuronal tangles, amyloid plaques, synaptic loss, and neuronal death – are not well understood (Hüll et al. 2000).

9.2 Components of Neuroinflammation in the Central Nervous System

Besides a frank inflammation with the immigration of blood-borne macrophages, B, and T cells into the central nervous system (CNS) in acute inflammatory diseases, a low-grade chronic neuroinflammatory reaction can be initiated and propagated within the CNS in the absence of immigration of leucocytes. Nearly all resident cells of the CNS are capable of synthesising pro-inflammatory cytokines, inflammation-associated enzymes, acute phase proteins and complement proteins, which may act in a tissue-specific and tissue restrictive neuroinflammatory reaction. In the CNS, microglia can produce interleukin (IL)-1 and both microglia and astroglia can produce IL-6 (Del Bo et al. 1995; Fiebich et al. 1997; Van Wagoner et al. 1999). In addition, microglia, astroglia and neurons are able synthesise prostaglandins via the expression of cyclooxygenase-2 (COX-2) (Yamagata et al. 1993; O'Banion et al. 1996a; Bauer et al. 1997; Hirst et al. 1999). Furthermore, all three types of cells are capable of synthesising some components of the complement cascade or acute phase proteins (Yasojima et al. 1999).

9.3 Microglial Activation in Alzheimer's Disease

In AD, several components of a neuroinflammatory reaction are activated in the CNS (Akiyama et al. 2000). Since the earliest description of AD at the beginning of the last century, an activation of glial cells and especially of small glial cells (microglia) has been consistently detected in the brains of demented patients (Lue et al. 1996). In non-demented elderly controls, high numbers of diffuse amyloid plaques may be present while neuritic amyloid plaques are nearly completely absent. In contrast to non-demented elderly controls, high numbers of microglial cells with an enlarged or phagocytotic morphology and expression of major histocompatibility complex class II are found in diffuse amyloid plaques in AD (Mackenzie 1994; Mackenzie et al. 1995; Sasaki et al. 1997; Sheng et al. 1997). Therefore, it was hypothesised that the activation of microglia may be the time-limiting step in the transition of a diffuse plaque into a neuritic plaque. Lewy body disease (LBD) may be the cause of dementia in 10% of all patients with cognitive decline and the second most common cause of dementia in the elderly (McKeith 2000). In LBD, diffuse but not neuritic amyloid plaques are commonly found in the cerebral cortex. In concordance with the role of inflammation in the formation of neuritic plaques, cortical microglial activation is not found in LBD (Rozemuller et al. 2000; Shepherd et al. 2000). Recently, further evidence has been provided from neuropathological studies that microglia are the driving force of fibrillary amyloid formation in AD (Wegiel et al. 2000).

There are several observations that contradict the view that microglial activation is only connected to Aβ pathology in AD. Neuropathological investigations show a widespread activation of microglia outside amyloid plaques in AD. High numbers of activated microglia cells are not only found in amyloid plaques but also in hippocampal areas without amyloid deposition (Roe et al. 1996; Sheng et al. 1997). Therefore, the deposition of Aβ peptides is neither a necessary nor a sufficient cause of microglial activation in AD. Aβ peptides, however, may facilitate microglial activation. In vitro components of amyloid plaques can stimulate microglial cells to increased neuronal toxicity (Giulian et al. 1996). Although the widespread neuronal cell death in AD is several times higher in amyloid plaques than in the surrounding tissue, most dying neurons are not associated with amyloid plaques (Lassmann et al. 1995; Sugaya et al. 1997). Microglial activation may be also connected to the formation of neurofibrillary tangles. In the axonal termination area of tangle bearing neurons, activated microglial cells are frequently found. The microglial activation in the hippocampus showed a much better correlation to the extent of tangle formation in the entorhinal cortex than to hippocampal amyloid deposition (DiPatre and Gelman 1997). In conclusion, in human autopsy tissue of AD, a massive microglial activation is found in and outside of amyloid deposits without signs of an effective mechanism of Aβ clearance by microglial cells, suggesting an overall harmful pro-inflammatory activation of glial cells.

9.4 Cytokines, Acute-Phase Proteins and Complement Components

Cytokines and acute-phase proteins are further signs of the neuroinflammatory activation in AD. The cytokines IL-1 and IL-6 have been found in amyloid plaques. IL-1 appears to occur early in plaque formation and is already expressed in microglia surrounding early, diffuse amyloid plaques (Mrak et al. 1995; Griffin et al. 1998). IL-1 activates astrocytes and induces the synthesis of IL-6 and a number of acute phase proteins such as α_1-antichymotrypsin (α_1-ACT) (Das and Potter 1995; Machein et al. 1995). In AD, α_1-ACT is found in amyloid plaques and may bind Aβ and modulate fibril formation (Abraham et al. 1988; Ma et al. 1996). Elevated levels of IL-6 are detectable in brain extracts of AD patients (Wood et al. 1993). IL-6 immunoreactivity has been found in both early diffuse amyloid plaques and neuritic amyloid plaques and may contribute to the progression of neuropathology (Hüll et al. 1995). In normal ageing and in AD the expression of IL-6 mRNA seems to be especially high in the vulnerable entorhinal brain areas (Luterman et al. 2000). Recently, it has been reported that polymorphisms of both IL-1 and IL-6 are associated with the development of AD (Papassotiropoulos et al. 1999; Du et al. 2000).

The synthesis of the complement component C1q is markedly elevated in AD (Yasojima et al. 1999). Binding of C1q to Aβ peptides may facilitate Aβ aggregation and binding of C1q to cell membranes (Valazquez et al. 1997).

9.5 Cyclooxygenase 1/2 and the Production of Prostaglandins

The key enzyme of prostaglandin synthesis, COX, exists in two homologous isoforms with divergent mechanisms of gene regulation and expression patterns. While earlier studies focused mainly on the role of the inducible COX-2 in AD, recent reports suggest that the "constitutively" expressed COX-1 also plays an important role in AD. Both COX-1 and COX-2 mRNA seem to be elevated in the frontal cortex in AD (Pasinetti and Aisen 1998). A well-controlled post-mortem study indicates a higher variability of COX-2 mRNA in the brains of AD patients as compared to age matched controls (Lukiw and Bazan 1997). Furthermore, a correlation between the presence of the transcription factor NF-κB in the cell nucleus and the level of COX-2 mRNA was found in brain tissues of AD patients and age-matched controls, suggesting that NF-κB is involved in the induction of COX-2 in the human brain (Lukiw and Bazan 1998).

COX-2 is expressed in neurons in the brains of AD patients as well as age-matched controls (O'Banion et al. 1996b). In neurons, COX-2 mRNA expression is rapidly regulated by synaptic activity and the COX-2 gene belongs to the immediate early genes. The physiological function of the highly regulated COX-2 expression in neurons is unclear. The enzymatic generation of oxygen radicals by COX-2 inside neurons adds to the amount of free radical load, which might lead to neurodegeneration. Divergent reports about an elevated expression of COX-2 in neurons in AD exists (O'Banion et al. 1996b ; Oka and Takashima 1997; Ho et al. 1999). However, considering the loss of synaptic transmission in dementia, neuronal COX-2 expression seems to be higher than one would expect.

In cultured astrocytes and rat microglial cells the expression of COX-2 with subsequent synthesis of prostaglandin E_2 (PGE_2) can be induced by various stimuli (O'Banion et al. 1996). On the other hand, PGE_2 is able to induce COX-2 in microglial cells (Minghetti et al. 1997). Therefore, some sort of autocrine or paracrine amplification of the COX-2 induction in microglial cells or a spreading of COX-2 expression between neurons and microglial cells seem to be possible. In animal models of hypoxia and in human stroke, COX-2 expression in microglial cells, astrocytes and neurons is markedly upregulated (Sanz et al.

1997; Sairanen et al. 1998). However, in contrast to the expression of COX-2 in neurons, the expression of COX-2 in microglia or astrocytes has not been convincingly shown in neuropathological specimen in AD. Given the fact that prostaglandins can modulate microglial function, the effects of paracrine- or autocrine-secreted prostaglandins on the reaction of microglia towards amyloid depositions should be investigated. In a neuropathological study, the activation of microglial cells in amyloid plaques in the brains of long-term users of nonsteroidal anti-inflammatory drugs (NSAIDs) was markedly reduced as compared to age-matched controls (Mackenzie and Munoz 1998).

Besides the elevated expression of COX-2 in AD, higher levels of prostaglandins are found in the cerebrospinal fluid (Oka and Takashima 1997; Pasinetti and Aisen 1998; Ho et al. 1999; Montine et al. 1999). However, besides COX-2, COX-1 expression may also be elevated in AD, and enzymatic activity of COX-1 may also be a possible source of prostaglandins in AD (Kitamura et al. 1999; Yermakova et al. 1999; Hoozemans et al. 2001). If COX-1 plays a major role in the production of prostaglandins in AD, new strategies focusing on selective COX-2 inhibitors may show only limited effects.

9.6 Neuroinflammation in Animal Models and Cell Cultures

In rats, intraventricular infusion of Aβ leads to an activation of microglial cells, which can be attenuated by NSAIDs (Netland et al. 1998). In several strains of transgenic mice, overexpression of a mutated human amyloid-precursor protein (APP) leads to the activation of microglia (Frautschy et al. 1998). Activation of microglial cells may be therefore stimulated directly by extracellular deposition of Aβ. Aβ is able to induce cytokine synthesis in cerebral cell cultures and therefore initiate a propagation of neuroinflammatory activation from microglia to astroglia and vice versa (Del Bo et al. 1995).

However, activation of neuroinflammation by Aβ may not be an unidirectional way of interaction. Chronic neuroinflammation itself may foster neurodegeneration. In rats, chronic ventricular infusion of lipopolysaccharides leads to microglial activation, enhanced processing of the amyloid precursor protein and neurodegeneration which is accentuated in the temporal lobe and the hippocampus (Hauss-Wegrzyniak et al. 1998). In this model, microglial activation persisted for a long time after cessation of lipopolysaccharide infusion (Hauss-Wegrzyniak et al. 2000). Furthermore, neuroinflammation lead to a loss of cholinergic neurons which was reduced by treatment with NSAIDs (Wenk et al. 2000). Therefore, neuroinflammation may be a self-propagating process which adds to the progression neuropathology. Recently, the anti-inflammatory drug ibuprofen has been shown to decrease the number of activated microglial cells in APP transgenic animals and moreover to reduce amyloid deposition and the numbers of dystrophic neurites (Lim et al. 2000). This finding strongly argues for a positive feedback loop between microglial activation and amyloid deposition.

IL-6 has been implicated in the neuroinflammatory cascade in AD (Hüll et al. 1996). IL-6 is one of the most potent cytokines to stimulate astrogliosis after intracerebral injection (Balasingam et al. 1994; Chiang et al. 1994). Transgenic animals with a brain-specific overexpression of IL-6 show morphological and functional alterations, such as a reduction of dendritic arborisation, loss of cholinergic hippocampal innervation, elevated synthesis of complement proteins, astrogliosis and deficits in long-term potentiation and memory function (Campbell et al. 1993; Steffenson et al. 1994; Bellinger et al. 1995; Heyser et al. 1997). Astrogliosis in AD has mostly been regarded as a final consequence of tissue destruction; however, astroglial activation may persist for a long time in AD, and activated astrocytes may play a more active part in the progression of neuropathology (Le Prince et al. 1993; Licastro et al. 1998).

9.7 Vaccination Studies in Animal Models

Vaccination with fibrillar human $A\beta_{1-42}$ has been shown to reduce the amount of $A\beta$ deposition in three different strains of transgenic mice overexpressing a mutant human form of the amyloid precursor protein (Schenk et al. 1999; Janus et al. 2000; Morgan et al. 2000). Moreover, a reduced cognitive decline has been shown for two strains after vaccination (Janus et al. 2000; Morgan et al. 2000). The effects of vaccination have been compared to the effects of the passive transfer of antibodies against $A\beta_{1-42}$, which may have similar effects (Bard et al. 2000). Currently a model is suggested that antibodies against $A\beta_{1-42}$ cross the blood–brain barrier and allow microglial cells to phagocytose $A\beta_{1-42}$ via binding of $A\beta_{1-42}$ antibodies to the microglial Fc-receptor (Bard et al. 2000). All in all, microglial activation has been shown to be reduced in vaccinated animals or in animals with topical application of $A\beta_{1-42}$ antibodies (Bacskai et al. 2001). However, whether the improved performance of vaccinated animals correlates with the reduction of microglial activation has not been calculated. Given the fact that in non-demented humans amyloid deposits are found without signs of tissue damage, the pure removal of amyloid deposition may not be sufficient to reduce cognitive impairment.

9.8 Epidemiological and Clinical Data on Anti-Inflammatory Drugs

One consistent finding in more than 14 epidemiological studies is a reduced risk of developing AD in subjects treated with anti-inflammatory drugs (Breitner 1996). In a neuropathological study, the activation of microglial cells in the brains of long-term users of NSAIDs was markedly reduced as compared to age-matched controls (Mackenzie and Munoz 1998). In 1993, a small clinical trial with 28 patients using indomethacin showed a reduced progression of AD (Rogers et al. 1993). Another clinical trial in 1999 with a combination of diclofenac and misoprostol showed only a trend for a reduced progression (Scharf et al. 1999). However, the statistical power of both pilot studies was too weak to draw final conclusions because of a high rate of dropouts due to gastro-intestinal disturbances. Only 12 of 24 patients with the combination of diclofenac and misoprostol completed the study. Another approach to modulate inflammatory activation in the CNS in AD has been the use of low-dose prednisone. A daily dose of 10 mg of prednisone given for 1 year did not reduce the progression of AD (Aisen et al. 2000). Different arguments have been raised to explain this result including the dose-depend lack of 10 mg prednisolone to influence CNS inflammation or specific side-effects of steroids on hippocampal degeneration (Hüll et al. 2000). Other anti-inflammatory steroidal agents with lower side-effects compared to glucocorticoids may be suitable for further investigation. Estrogen derivates are currently investigated; however, results from clinical trials have not shown beneficial effects up to now (Henderson et al. 2000; Mulnard et al. 2000).

Recently two independent retrospective epidemiological studies suggested a protective role of statins against the development of dementia (Jick et al. 2000; Wolozin et al. 2000). Statins are commonly regarded as lipid-lowering agents acting mainly as 3-hydroxy-3-methylglutaryl coenzyme A reductase inhibitors. However, recent papers suggest that statins reduce proinflammatory activation and IL-6 synthesis independent of their lipid-lowering effect (Rosenson 1999; Grip et al. 2000; Palinski 2001). Further trials are now on their way to explore the potential role of statins in AD.

9.9 Neuroinflammation and Possible Drug Targets

The approaches to reduce inflammation by inhibition of COX have been hampered by the side-effects of non selective COX inhibitors. Therefore, ongoing clinical trials with anti-inflammatory drugs focus on selective COX-2 inhibitor such as celecoxib or rofecoxib, which show only mild gastro-intestinal side-effects (Geis 1999). Observations in cell cultures and transgenic animals suggest a role of COX-2 in neuronal cell death (Ho et al. 1998; Kelley et al. 1999). However, the approach to suppress inflammatory activation in AD by selective inhibition of COX-2 may be not sufficient. First of all, recent data point to the role of COX-1 in the cerebral production of prostaglandins, which is not suppressed by selective COX-2 inhibitors (Yermakova et al. 1999; Hoozemans et al. 2001). Second, the epidemiological data concerning the reduced risk of developing AD by the use of NSAIDs combine the observation made with different NSAIDs which are not only non-selective COX-inhibitors but may have also other anti-inflammatory properties. Third, different cell culture or animals studies to evaluate the potential of NSAIDs to suppress inflammation and the clinical trial of Rogers et al. (1993) used indomethacin or ibuprofen, which are known as agonists of peroxisome proliferator-activated receptor-gamma (PPARγ) besides their effects on COX-1 and COX-2. Results of the ongoing trails with selective COX-2 inhibitors will presumably be available within the next year.

The PPARγ belongs to the nuclear receptor superfamily of steroid and retinoid receptors (Pershadsingh 1999). The expression of PPARγ is elevated in AD (Kitamura et al. 1999). NSAIDs, such as indomethacin or ibuprofen, do not only inhibit the enzymatic activity of COX-2 but modulate also gene transcription by agonistic binding to PPARγ (Lehmann et al. 1997). Furthermore, the class of thiazolidinedione antidiabetic drugs such as troglitazone, rosiglitazone or pioglitazone are agonists of PPARγ (Camp et al. 2000). In several neural cell culture models and animal models, PPARγ agonist have been shown to reduce the expression of COX-2, IL-6, inducible nitric-oxide synthase and neuronal damage (Combs et al. 2000; Heneka et al. 2000; Ogawa et al. 2000). PPARγ agonist also reduced proinflammatory activation of microglial cells stimulated with Aβ (Combs et al. 2000).

In microglial cells, Aβ peptides activate p38 mitogen-activated protein kinase (p38 MAPK), induce the phosphorylation of the transcription factor cAMP-response element binding protein (CREB) and stimulate the production of superoxide by activation of protein tyrosine kinases (McDonald et al. 1997; McDonald et al. 1998). In vitro components of amyloid plaques can stimulate microglial cells to increase neuronal toxicity (Giulian et al. 1996). Besides the role of p38 MAPK in Aβ activation of microglial cells, activation of p38 MAPK has also been shown in neurons in neuropathological studies of AD (Arendt et al. 1995; Hensley et al. 1999). Furthermore, the enzymatic activity of p38 MAPK has also been linked to the formation of hyperphosphorylated tau and the expression of COX-2 in neurons (Goedert et al. 1997; Fiebich et al. 2000). Therefore, inhibition of p38MAPK may reduce the progression of neuropathology. Different pyridinylimidazole derivates such as SB203580 have been shown to inhibit selectively the catalytic activity of p38 MAPK and reduce inflammation in animal models of arthritis, septic shock and myocardial injury (Lee et al. 2000). Another inhibitor of p38 MAPK, SB242235, has been shown to reduce the progression of inflammation and the expression of IL-6 in rats (Badger et al. 2000).

Although the role of IL-6 in the CNS is complex, inhibition of the elevated IL-6 synthesis in AD may be another target for disease modifying drugs (Hüll et al. 1996; Gruol and Nelson 1997). Both, p38MAPK inhibitors and PPARγ agonists reduce IL-6 synthesis (Badger et al. 2000; Combs et al. 2000). COX inhibitors may reduce PGE_2 release which stimulate IL-6 synthesis in astrocytes (Fiebich et al. 1997). Therefore, all of these three strategies may already act on the IL-6 pathway of neuroinflammation. Furthermore, two novel NSAIDs, tepoxalin and

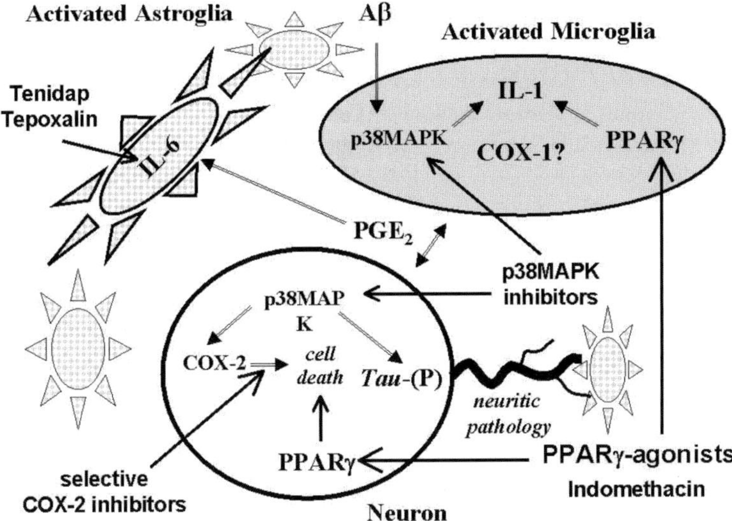

Fig. 1. Pathways of neuroinflammation and potential drug targets. Aβ and a further unknown process activate microglial cells in AD which synthesize IL-1 and stimulate astroglial IL-6 production. PGE_2 produced by neurons and microglial cells may act as a paracrine modulator of neuroinflammation. Inhibitors of p38MAPK or agonists of PPARγ, such as indomethacin, may reduce COX-2 or IL-1 expression and neuronal cell death. Additionally, inhibitors of p38MAPK may also reduce tau pathology. Newly developed NSAIDs such as tenidap or tepoxalin may reduce astroglial IL-6 expression and hence astrogliosis. Selective inhibition of neuronal COX-2 may reduce neuronal cell death

tenidap, have been shown to reduce IL-6 synthesis in neural cells independent of COX inhibition (Fiebich et al. 1996, 1999). Figure 1 gives a summary of the different targets that interfere with the neuroinflammatory circuits in AD.

9.10 Consideration for Further Research

The reduction of neuroinflammation may be a suitable target for drugs with the aim to protect against the development of dementia in AD. Several drugs with good epidemiological evidence to reduce the risk to develop AD have failed in clinical trial with moderately demented patients. Therefore, different neuropathological mechanisms may prevail in different phases of the continuous progression of CNS damage in AD. Despite advances in cerebrospinal fluid analysis and neuroimaging (Hampel et al. 1997, 1998, 2001) the clinical diagnosis of AD still depends on the presence of a clear cognitive dysfunction which might already signify a loss of 50% of synapses in certain cortical areas. At the point of clinically obvious dementia, minor additive structural damage may lead to major functional worsening if the redundancy of the synaptic connectivity is exhausted. Therefore drugs which cannot demonstrate efficacy in the treatment of clinically obvious dementia may be still valuable in the prevention of early tissue damage in the clinically asymptomatic early phase of AD.

The ideal drug against neuroinflammation in AD will have a good penetrance of the blood–brain barrier, reduce microglial activation but does not compromise microglial Aβ clearance, reduce astroglial activation, and protect neurons from cell death. Although it may not be possible to find a drug which combines all these attributes in a single substance, inhibitors of cyclooxygenase with additional features such as blockage of cytokine synthesis or activation of PPARγ are already available. Trials with these substances will have to be conducted for several years to demonstrate a disease-altering effect in already slightly demented patients. Well tolerable anti-inflammatory drugs may also become a good choice for clinical trials to prevent cognitive decline in elderly people with mild cognitive impairment.

Acknowledgements. M. Hüll is grateful for the support by the Alzheimer Forschung Initiative e. V.

References

Abraham CR, Selkoe DJ, Potter H (1988) Immunochemical identification of the serine protease inhibitor alpha 1-antichymotrypsin in the brain amyloid deposits of Alzheimer's disease. Cell 52:487–501

Aisen PS, Davis KL, Berg JD, Schafer K, Campbell K, et al (2000) A randomized controlled trial of prednisone in Alzheimer's disease. Alzheimer's Disease Cooperative Study. Neurology 54:588–593

Akiyama H, Barger S, Barnum S, Bradt B, Bauer J, et al (2000) Inflammation and Alzheimer's disease. Neurobiol Aging 21:383–421

Arendt T, Holzer M, Grossmann A, Zedlick D, Bruckner MK (1995) Increased expression and subcellular translocation of the mitogen activated protein kinase kinase and mitogen-activated protein kinase in Alzheimer's disease. Neuroscience 68:5–18

Bacskai BJ, Kajdasz ST, Christie RH, Carter C, Games D, et al (2001) Imaging of amyloid-beta deposits in brains of living mice permits direct observation of clearance of plaques with immunotherapy. Nat Med 7:369–372

Badger AM, Griswold DE, Kapadia R, Blake S, Swift BA, et al (2000) Disease-modifying activity of SB 242235, a selective inhibitor of p38 mitogen-activated protein kinase, in rat adjuvant-induced arthritis. Arthritis Rheum 43:175–183

Balasingam V, Tejada-Berges T, Wright E, Bouckova R, Yong VW (1994) Reactive astrogliosis in the neonatal mouse brain and its modulation by cytokines. J Neurosci 14:846–856

Bard F, Cannon C, Barbour R, Burke RL, Games D, et al (2000) Peripherally administered antibodies against amyloid beta-peptide enter the central nervous system and reduce pathology in a mouse model of Alzheimer disease. Nat Med 6:916–919

Bauer MKA, Lieb K, Schulze-Osthoff K, Berger M, Gebicke-Haerter PJ, et al (1997) Expression and regulation of cyclooxygenase-2 in rat microglia. Eur J Biochem 243:726–731

Bellinger FP, Madamba SG, Campbell IL, Siggins GR (1995) Reduced long-term potentiation in the dentate gyrus of transgenic mice with cerebral overexpression of interleukin-6. Neurosci Lett 198:95–98

Breitner JCS (1996) Inflammatory processes and antiinflammatory drugs in Alzheimer's disease: A current appraisal. Neurobiol Aging 17:789–794

Camp HS, Li O, Wise SC, Hong YH, Frankowski CL, et al (2000) Differential activation of peroxisome proliferator-activated receptor- gamma by troglitazone and rosiglitazone. Diabetes 49:539–547

Campbell IL, Abraham CR, Masliah E, Kemper P, Inglis JD, et al (1993) Neurologic disease induced in transgenic mice by cerebral overexpression of interleukin 6. Proc Natl Acad Sci USA 90:10061–10065

Chiang C-S, Stalder A, Samimi A, Campbell I (1994) Reactive gliosis as a consequence of interleukin-6 expression in the brain: studies in transgenic mice. Dev Neurosci 16:212–221

Combs CK, Johnson DE, Karlo JC, Cannady SB, Landreth GE (2000) Inflammatory mechanisms in Alzheimer's disease: inhibition of beta- amyloid-stimulated proinflammatory responses and neurotoxicity by PPARgamma agonists. J Neurosci 20:558–567

Das S, Potter H (1995) Expression of the Alzheimer amyloid-promoting factor antichymotrypsin is induced in human astrocytes by IL-1. Neuron 14:447–456

Del Bo R, Angeretti N, Lucca E, Grazia De Simoni M, Forloni G (1995) Reciprocal control of inflammatory cytokines, IL-1 and IL- 6, and b-amyloid production in cultures. Neurosci Lett 188:70–74

DiPatre PL, Gelman BB (1997) Microglial cell activation in aging and Alzheimer disease: Partial linkage with neurofibrillary tangle burden in the hippocampus. J Neuropathol Exp Neurol 56:143–149

Du Y, Dodel RC, Eastwood BJ, Bales KR, Gao F, et al (2000) Association of an interleukin 1 alpha polymorphism with Alzheimer's disease. Neurology 55:480–483

Fiebich BL, Lieb K, Hüll M, Berger M, Bauer J (1996) Effects of NSAIDs on IL-1 beta-induced IL-6 mRNA and protein synthesis in human astrocytoma cells. Neuroreport 7:1209–1213

Fiebich BL, Hüll M, Lieb K, Gyufko K, Berger M, et al (1997) Prostaglandin E2 induces interleukin-6 synthesis in human astrocytoma cells. J Neurochem 68:704–709

Fiebich BL, Hofer TJ, Lieb K, Huell M, Butcher RD, et al (1999) The non-steroidal anti-inflammatory drug tepoxalin inhibits interleukin- 6 and alpha1-anti-chymotrypsin synthesis in astrocytes by preventing degradation of IkappaB-alpha. Neuropharmacology 38:1325–1333

Fiebich BL, Mueksch B, Boehringer M, Hüll M (2000) IL-1 beta induces cyclooxygenase-2 and prostaglandin E2 synthesis in human neuroblastoma cells: Involvement of p38 mitogen activated protein kinase and nuclear factor-kappa B. J Neurochem 75:2020–2028

Frautschy SA, Yang F, Irrizarry M, Hyman B, Saido TC, et al (1998) Microglial response to amyloid plaques in APPsw transgenic mice. Am J Pathol 152:307–317

Geis GS (1999) Update on clinical developments with celecoxib, a new specific COX-2 inhibitor: what can we expect? J Rheumatol 26 [Suppl] 56:31–36

Giulian D, Haverkamp LJ, Yu JH, Karshin W, Tom D, et al (1996) Specific domains of beta-amyloid from Alzheimer plaque elicit neuron killing in human microglia. J Neurosci 16:6021–6037

Goedert M, Hasegawa M, Jakes R, Lawler S, Cuenda A, et al (1997) Phosphorylation of microtubule-associated protein tau by stress-activated protein kinases. FEBS Lett 409:57–62

Griffin WS, Sheng JG, Royston MC, Gentleman SM, McKenzie JE, et al (1998) Glial-neuronal interactions in Alzheimer's disease: the potential role of a 'cytokine cycle' in disease progression. Brain Pathol 8:65–72

Grip O, Janciauskiene S, Lindgren S (2000) Pravastatin down-regulates inflammatory mediators in human monocytes in vitro. Eur J Pharmacol 410:83–92

Gruol DL, Nelson TE (1997) Physiological and pathological roles of interleukin-6 in the central nervous system. Mol Neurobiol 15:307–339

Hampel H, Teipel SJ, Kotter HU, Horwitz B, Pfluger T, et al (1997) Structural magnetic resonance tomography in diagnosis and research of Alzheimer type dementia. Nervenarzt 68:365–378

Hampel H, Sunderland T, Kotter HU, Schneider C, Teipel, et al (1998) Decreased soluble interleukin-6 receptor in cerebrospinal fluid of patients with Alzheimer's disease. Brain Res 780:356–359

Hampel H, Buerger K, Kohnken R, Teipel SJ, Zinkowski R, et al (2001) Tracking of Alzheimer's disease progression with cerebrospinal fluid tau protein phosphorylated at threonine 231. Ann Neurol 49:545–546

Hauss-Wegrzyniak B, Dobrzanski P, Stoehr JD, Wenk GL (1998) Chronic neuroinflammation in rats reproduces components of the neurobiology of Alzheimer's disease. Brain Res 780:294–303

Hauss-Wegrzyniak B, Vraniak PD, Wenk GL (2000) LPS-induced neuroinflammatory effects do not recover with time. Neuroreport 11:1759–1763

Henderson VW, Paganini-Hill A, Miller BL, Elble RJ, Reyes PF, et al (2000) Estrogen for Alzheimer's disease in women: randomized, double-blind, placebo-controlled trial. Neurology 54:295–301

Heneka MT, Klockgether T, Feinstein DL (2000) Peroxisome proliferator-activated receptor-gamma ligands reduce neuronal inducible nitric oxide synthase expression and cell death in vivo. J Neurosci 20:6862–6867

Hensley K, Floyd RA, Zheng NY, Nael R, Robinson KA, et al (1999) p38 kinase is activated in the Alzheimer's disease brain. J Neurochem 72:2053–2058

Heyser CJ, Masliah E, Samimi A, Campbell IL, Gold LH (1997) Progressive decline in avoidance learning paralleled by inflammatory neurodegeneration in transgenic mice expressing interleukin 6 in the brain. Proc Natl Acad Sci USA 94:1500–1505

Hirst WD, Young KA, Newton R, Allport VC, Marriott DR, et al (1999) Expression of COX-2 by normal and reactive astrocytes in the adult rat central nervous system. Mol Cell Neurosci 13:57–68

Ho L, Osaka H, Aisen PS, Pasinetti GM (1998) Induction of cyclooxygenase (COX)-2 but not COX-1 gene expression in apoptotic cell death. J Neuro Immuno 89:142–149

Ho L, Pieroni C, Winger D, Purohit DP, Aisen PS, et al (1999) Regional distribution of cyclooxygenase-2 in the hippocampal formation in Alzheimer's disease. J Neurosci Res 57:295–303

Hoozemans JJ, Rozemuller A, Janssen I, De Groot CJ, Veerhuis R, et al (2001) Cyclooxygenase expression in microglia and neurons in Alzheimer's disease and control brain. Acta Neuropathol (Berl) 101:2–8

Hüll M, Lieb K, Fiebich BL (2000) Anti-inflammatory drugs: a hope for Alzheimer's disease? Exp Opin Invest Drugs 9:671–683

Hüll M, Strauss S, Volk B, Berger M, Bauer J (1995) Interleukin-6 is present in early stages of plaque formation and is restricted to the brains of Alzheimer's disease patients. Acta Neuropathol (Berl) 89:544–551

Hüll M, Fiebich BL, Lieb K, Strauss S, Berger M, et al (1996) Interleukin-6-associated inflammatory processes in Alzheimer's disease: New therapeutic options. Neurobiol Aging 17:795–800

Janus C, Pearson J, McLaurin J, Mathews PM, Jiang Y, et al (2000) A beta peptide immunization reduces behavioural impairment and plaques in a model of Alzheimer's disease. Nature 408:979–982

Jick H, Zornberg GL, Jick SS, Seshadri S, Drachman DA (2000) Statins and the risk of dementia. Lancet 356:1627–1631

Kelley KA, Ho L, Winger D, Freire-Moar J, Borelli CB, et al (1999) Potentiation of excitotoxicity in transgenic mice overexpressing neuronal cyclooxygenase-2. Am J Pathol 155:995–1004

Kitamura Y, Shimohama S, Koike H, Kakimura J, Matsuoka Y, et al (1999) Increased expression of cyclooxygenases and peroxisome proliferator-activated receptor-gamma in Alzheimer's disease brains. Biochemical & Biophysical Research Communications 254:582–586

Lassmann H, Bancher C, Breitschopf H, Wegiel J, Bobinski M, et al (1995) Cell death in Alzheimer's disease evaluated by DNA fragmentation in situ. Acta Neuropathol (Berl) 89:35–41

Le Prince G, Delaere P, Fages C, Duyckaerts C, Hauw JJ, et al (1993) Alterations of glial fibrillary acidic protein mRNA level in the aging brain and in senile dementia of the Alzheimer type. Neurosci Lett 151:71–73

Lee JC, Kumar S, Griswold DE, Underwood DC, Votta BJ, et al (2000) Inhibition of p38 MAP kinase as a therapeutic strategy. Immunopharmacology 47:185–201

Lehmann JM, Lenhard JM, Oliver BB, Ringold GM, Kliewer SA (1997) Peroxisome proliferator-activated receptors alpha and gamma are activated by indomethacin and other non-steroidal anti-inflammatory drugs. J biol Chem 272:3406–3410

Licastro F, Mallory M, Hansen LA, Masliah E (1998) Increased levels of alpha-1-antichymotrypsin in brains of patients with Alzheimer's disease correlate with activated astrocytes and are affected by APOE 4 genotype. J Neuro Immuno 88:105–110

Lim GP, Yang F, Chu T, Chen P, Beech W, et al (2000) Ibuprofen suppresses plaque pathology and inflammation in a mouse model for Alzheimer's disease. J Neurosci 20:5709–5714

Lue LF, Brachova L, Civin WH, Rogers J (1996) Inflammation, Ab deposition, and neurofibrillary tangle formation as correlates of Alzheimer's disease neurodegeneration. J Neuropathol Exp Neurol 55:1083–1088

Lukiw WJ, Bazan NG (1997) Cyclooxygenase 2 RNA message abundance, stability, and hypervariability in sporadic Alzheimer neocortex. J Neurosci Res 50:937–945

Lukiw WJ, Bazan NG (1998) Strong nuclear factor-kappaB-DNA binding parallels cyclooxygenase-2 gene transcription in aging and in sporadic Alzheimer's disease superior temporal lobe neocortex. J Neurosci Res 53:583–592

Luterman JD, Haroutunian V, Yemul S, Ho L, Purohit D, et al (2000) Cytokine gene expression as a function of the clinical progression of Alzheimer disease dementia. Arch Neurol 57:1153–1160

Ma JY, Brewer HB, Jr., Potter H (1996) Alzheimer Ab neurotoxicity: Promotion by antichymotrypsin, ApoE4; Inhibition by Ab-related peptides. Neurobiol Aging 17:773–780

Machein U, Lieb K, Hüll M, Fiebich BL (1995) IL-1 beta and TNF alpha, but not IL-6, induce alpha 1-antichymotrypsin expression in the human astrocytoma cell line U373 MG. Neuroreport 6:2283–2286

Mackenzie IR (1994) Senile plaques do not progressively accumulate with normal aging. Acta Neuropathol Berl 87:520–525

Mackenzie IR, Munoz DG (1998) Nonsteroidal anti-inflammatory drug use and Alzheimer-type pathology in aging. Neurology 50:986–990

Mackenzie IR, Hao CH, Munoz DG (1995) Role of microglia in senile plaque formation. Neurobiol Aging 16:797–804

McDonald DR, Brunden KR, Landreth GE (1997) Amyloid fibrils activate tyrosine kinase-dependent signaling and superoxide production in microglia. J Neurosci 17:2284–2294

McDonald DR, Bamberger ME, Combs CK, Landreth GE (1998) beta-Amyloid fibrils activate parallel mitogen-activated protein kinase pathways in microglia and THP1 monocytes. J Neurosci 18:4451–4460

McKeith IG (2000) Spectrum of Parkinson's disease, Parkinson's dementia, and lewy body dementia. Neurol Clin 18:865–902

Minghetti L, Polazzi E, Nicolini A, Créminon C, Levi G (1997) Up-regulation of cyclooxygenase-2 expression in cultured microglia by prostaglandin E_2, cyclic AMP and non-steroidal anti-inflammatory drugs. Eur J Neurosci 9:934–940

Montine TJ, Sidell KR, Crews BC, Markesbery WR, Marnett LJ, et al (1999) Elevated CSF prostaglandin E2 levels in patients with probable AD. Neurology 53:1495–1498

Morgan D, Diamond DM, Gottschall PE, Ugen KE, Dickey C, et al (2000) A beta peptide vaccination prevents memory loss in an animal model of Alzheimer's disease. Nature 408:982–985

Mrak RE, Sheng JG, Griffin WS (1995) Glial cytokines in Alzheimer's disease: review and pathogenic implications. Hum Pathol 26(8):816–823

Mulnard RA, Cotman CW, Kawas C, Van Dyck CH, Sano M, et al (2000) Estrogen replacement therapy for treatment of mild to moderate Alzheimer disease: a randomized controlled trial. Alzheimer's Disease Cooperative Study. JAMA 283:1007–1015

Netland EE, Newton JL, Majocha RE, Tate BA (1998) Indomethacin reverses the microglial response to amyloid beta-protein. Neurobiol Aging 19:201–204

O'Banion MK, Chang JW, Coleman PD (1996b) Decreased expression of prostaglandin G/H synthase-2 (PGHS-2) in Alzheimer's disease brain. Adv Exp Med Biol 407:171–177

O'Banion MK, Miller JC, Chang JW, Kaplan MD, Coleman PD (1996a) Interleukin-1b induces prostaglandin G/H synthase-2 (cyclooxygenase-2) in primary murine astrocyte cultures. J Neurochem 66:2532–2540

Ogawa O, Umegaki H, Sumi D, Hayashi T, Nakamura A, et al (2000) Inhibition of inducible nitric oxide synthase gene expression by indomethacin or ibuprofen in beta-amyloid protein-stimulated J774 cells. Eur J Pharmacol 408:137–141

Oka A, Takashima S (1997) Induction of cyclo-oxygenase 2 in brains of patients with Down's syndrome and dementia of Alzheimer type: Specific localization in affected neurones and axons. Neuroreport 8:1161–1164

Palinski W (2001) New evidence for beneficial effects of statins unrelated to lipid lowering. Arterioscler Thromb Vasc Biol 21:3–5

Papassotiropoulos A, Bagli M, Jessen F, Bayer TA, Maier W, et al (1999) A genetic variation of the inflammatory cytokine interleukin-6 delays the initial onset and reduces the risk for sporadic Alzheimer's disease. Ann Neurol 45:666–668

Pasinetti GM, Aisen PS (1998) Cyclooxygenase-2 expression is increased in frontal cortex of Alzheimer's disease brain. Neuroscience 87:319–324

Pershadsingh HA (1999) Pharmacological peroxisome proliferator-activated receptor gamma ligands: emerging clinical indications beyond diabetes. Exp Opin Invest Drugs 8:1859–1872

Roe MT, Dawson DV, Hulette CM, Einstein G, Crain BJ (1996) Microglia are not exclusively associated with plaque-rich regions of the dentate gyrus in Alzheimer's disease. J Neuropathol Exp Neurol 55:366–371

Rogers J, Kirby LC, Hempelman SR, Berry DL, McGeer PL, et al (1993) Clinical trial of indomethacin in Alzheimer's disease. Neurology 43:1609–1611

Rosenson RS (1999) Non-lipid-lowering effects of statins on atherosclerosis. Curr Cardiol Rep 1:225–232

Rozemuller AJ, Eikelenboom P, Theeuwes JW, Jansen Steur EN, de Vos RA (2000) Activated microglial cells and complement factors are unrelated to cortical Lewy bodies. Acta Neuropathol (Berl) 100:701–708

Sairanen T, Ristimaki A, Karjalainen-Lindsberg ML, Paetau A, Kaste M, et al (1998) Cyclooxygenase-2 is induced globally in infarcted human brain. Ann Neurol 43:738–747

Sanz O, Estrada A, Ferrer I, Planas AM (1997) Differential cellular distribution and dynamics of HSP70, cyclooxygenase-2, and c-Fos in the rat brain after transient focal ischemia or kainic acid. Neuroscience 80:221–232

Sasaki A, Yamaguchi H, Ogawa A, Sugihara S, Nakazato Y (1997) Microglial activation in early stages of amyloid b protein deposition. Acta Neuropathol (Berl) 94:316–322

Scharf S, Mander A, Ugoni A, Vajda F, Christophidis N (1999) A double-blind, placebo-controlled trial of diclofenac/misoprostol in Alzheimer's disease. Neurology 53:197–201

Schenk D, Barbour R, Dunn W, Gordon G, Grajeda H, et al (1999) Immunization with amyloid-beta attenuates Alzheimer-disease-like pathology in the PDAPP mouse. Nature 400:173–177

Sheng JG, Mrak RE, Griffin WST (1997) Neuritic plaque evolution in Alzheimer's disease is accompanied by transition of activated microglia from primed to enlarged to phagocytic forms. Acta Neuropathol (Berl) 94:1–5

Shepherd CE, Thiel E, McCann H, Harding AJ, Halliday GM (2000) Cortical inflammation in Alzheimer disease but not dementia with Lewy bodies. Arch Neurol 57:817–822

Steffenson SC, Campbell IL, Henriksen SJ (1994) Site-specific hippocampal pathophysiology due to cerebral overexpression of interleukin-6 in transgenic mice. Brain Res 652:149–153

Sugaya K, Reeves M, McKinney M (1997) Topographic associations between DNA fragmentation and Alzheimer's disease neuropathology in the hippocampus. Neurochem Int 31:275–281

Valazquez P, Cribbs DH, Poulos TL, Tenner AJ (1997) Aspartate residue 7 in amyloid b-protein is critical for classical complement pathway activation: Implications for Alzheimer's disease pathogenesis. Nat Med 3:77–79

Van Wagoner NJ, Oh JW, Repovic P, Benveniste EN (1999) Interleukin-6 (IL-6) production by astrocytes: autocrine regulation by IL-6 and the soluble IL-6 receptor. J Neurosci 19:5236–5244

Wegiel J, Wang KC, Tarnawski M, Lach B (2000) Microglia cells are the driving force in fibrillar plaque formation, whereas astrocytes are a leading factor in plague degradation. Acta Neuropathol (Berl) 100:356–364

Wenk GL, McGann K, Mencarelli A, Hauss-Wegrzyniak B, Del Soldato P, et al (2000) Mechanisms to prevent the toxicity of chronic neuroinflammation on forebrain cholinergic neurons. Eur J Pharmacol 402:77–85

Wolozin B, Kellman W, Ruosseau P, Celesia GG, Siegel G (2000) Decreased prevalence of Alzheimer disease associated with 3-hydroxy-3-methyglutaryl coenzyme A reductase inhibitors. Arch Neurol 57:1439–1443

Wood JA, Wood PL, Ryan R, Graff-Radford NR, Pilapil C, et al (1993) Cytokine indices in Alzheimer's temporal cortex: no changes in mature IL-1 beta or IL-1 RA but increases in the associated acute phase proteins IL-6, alpha-2-macroglobulin and C-reactive protein. Brain Res 629:245–252

Yamagata K, Andreasson KI, Kaufmann WE, Barnes CA, Worley PF (1993) Expression of a mitogen-inducible cyclooxygenase in brain neurons: regulation by synaptic activity and glucocorticoids. Neuron 11:371–386

Yasojima K, Schwab C, McGeer EG, McGeer, PL (1999) Up-regulated production and activation of the complement system in Alzheimer's disease brain. Am J Pathol 154:927–936

Yermakova AV, Rollins J, Callahan LM, Rogers J, O'Banion MK (1999) Cyclooxygenase-1 in human Alzheimer and control brain: quantitative analysis of expression by microglia and CA3 hippocampal neurons. J Neuropathol Exp Neurol 58:1135–1146

10 The Concept of *In Vivo* Imaging of Neuroinflammation with [^{11}C](*R*)-PK11195 PET

A. Cagnin, A. Gerhard, R.B. Banati

10.1 The PK11195 Binding Site 180
10.2 The Cellular Source of PK11195 Binding in the CNS 181
10.3 [^{11}C](*R*)-PK11195 PET Imaging and the Concept of "Neuroinflammation" 182
10.4 What Is the Significance of a Persistent "Neuroinflammatory" Response? 185
10.5 Potential Role of Microglia and the "Peripheral Benzodiazepine Binding Site" in Neuronal Regeneration 186
References 187

At present, most brain imaging approaches using specific radioisotopes and positron emission tomography (PET) are based on the known distribution of e.g. a neuroreceptor or metabolic pathway in the CNS of normal individuals, for example the normal signal pattern of [^{18}F] FDG (2-fluoro-2deoxy-D-glucose) in the healthy cortex. Against this normal standard, the pathological abnormalities in patients with brain disease are usually defined as deficits, i.e. relative losses of regional signal. There are some notable exceptions, such as in tumours, where the relevant information of the image lies in a regional signal increase. In neurodegenerative diseases, however, current neuroimaging with radiotracers mainly provides "deficit-images" that are more or less closely correlated to a clinically observed deficit in brain function, raising the occasional question as to what their added value for the understanding of the disease process might be. The number of in vivo probes that would allow us to detect and measure the de novo expression of molecules specifically associated with those cellular changes that characterise a neuropathological tissue reaction, i.e. the "positive phenomenology" of the disease, is still very limited.

In contrast, magnetic resonance imaging (MRI) can, depending on the specific spin-echo sequences, measure a large number of local signal perturbations in brain tissue affected by disease. The high sensitivity of MRI to detect disease-associated changes, however, is not matched by an equal degree of specificity. Thus, brain tissue with similar MR signal intensity may look histopathologically very different, e.g. neuronal vacuolation can be indistinguishable from dense astrogliosis (Chung et al. 1999). The aim of multi-modality imaging is, therefore, to combine the spatial resolution and sensitivity of MRI with the high molecular and cellular specificity of PET.

Here, we briefly outline the rational for employing [^{11}C](*R*)-PK11195 PET to detect activated microglia in vivo and use their presence as a measure of neuronal damage and disease activity. We regard its application as a cell biology-based "in vivo neuropathology". Consequently, the image information is best interpreted in the context of the cellular pathology of brain disease/damage rather then in the (often syndromal) definitions of clinical diagnosis.

10.1 The PK11195 Binding Site

The isoquinoline PK11195 was originally discovered as a compound that partially displaces certain benzodiazepines, such as diazepam, from a site that is structurally and functionally unrelated to the central benzodiazepine receptor associated with gamma-aminobutyric acid (GABA)-regulated channels. Particularly abundant in peripheral organs and haematogenous cells but barely present in the normal CNS, the binding site for PK11195 was named "peripheral benzodiazepine binding site" (PBBS) (for review see: Hertz 1993). Subsequently, the PBBS was found to co-precipitate with the outer membrane of mitochondria (Anholt 1986), hence its other name, mitochondrial benzodiazepine receptor (MBR). However, PK11195-binding is also present in non-mitochondrial fractions of brain extracts, and mitochondria-free erythrocytes (Olson et al. 1988; Hertz 1993) while immunocytochemical staining hints to the possible presence of PBBS in cell nuclei (Hardwick et al. 1999). Amongst others the PBBS plays an important role in steroid synthesis (Krueger and Papadopoulos 1993) and regulates immunological responses in mononuclear phagocytes (Pawlikowski 1993) The numerous other putative functions of the PBBS, that still have to merge into a coherent theory of its biological role, have recently been reviewed by Gavish et al. (1999).

10.2 The Cellular Source of PK11195 Binding in the CNS

In vitro, astrocytes are found to have high binding of PK11195 (Hertz 1993; Itzhak et al. 1996; Park et al. 1996). However, observations made in a number of experimental lesion models and diseases with blood–brain barrier damage suggested that focally increased PK11195 binding is due to binding to infiltrating haematogenous cells (Benavides et al. 1988; Dubois et al 1988). Subsequently, a number of in vivo studies reported that the distribution pattern of increased PBBS expression matched more closely the distribution of activated microglia than that of reactive astrocytes (Dubois et al. 1988; Myers et al. 1991; Stephenson et al 1995; Conway et al. 1998). Further direct evidence using axotomy models demonstrated (Banati et al. 1997) that activated microglia are the main source of PK11195 binding in vivo. In support of these findings, high-resolution microautoradiography with [^{3}H](*R*)-PK11195 combined with immunohistochemical cell identification performed on the same tissue section in inflammatory disease, i.e. multiple sclerosis (MS) and experimental autoimmune encephalomyelitis (EAE), has shown that increased binding of [^{3}H](*R*)-PK11195 is found on infiltrating blood-borne cells but also on activated microglia (Banati et al. 2000). The latter appear to become the dominant source of binding in areas without any obvious histopathology and remote from the primary pathological focus. Some discrepancy, however, still remains: in neurotoxic lesion model, immunoreactivity primarily in and around the nucleus of reactive hippocampal astrocytes was detected by a polyclonal antibody against the peripheral benzodiazepine receptor (Kuhlmann and

Guilarte 2000). The failure to find a complete match of the reported immunocytochemical stain for the peripheral benzodiazepine receptor with the cellular distribution of the microautoradiographic PK11195-label may either have technical reasons, such as sensitivity and specificity of the various detection methods, or indicate that the immunocytochemically detected PBBS is not completely identical with the autoradiographically detected PK11195 binding sites. At present, therefore, it may thus be advisable not to view (*R*)-PK11195 binding as synonymous with the PBBS.

With respect to above microautoradiographic double-labelling data (Banati et al. 2000), it is important that the relative cellular selectivity for activated microglia has been established using the R-enantiomer of PK11195, which has a higher affinity for the PK11195-binding site, rather than the commonly used racemate (Shah et al. 1994). The lack of significantly increased [^{11}C](*R*)-PK11195 binding in astrocyte-rich tissue, such as in patients with hippocampal sclerosis (Banati et al. 1999b) further supports the view that microglia are the dominant site of (*R*)-PK11195 binding in vivo. Importantly, these patients had a low seizure frequency, as one might expect frequent seizures to induce pathological changes with activation of microglia and consequently increases in PK11195-binding sites.

10.3 [^{11}C](*R*)-PK11195 PET Imaging and the Concept of "Neuroinflammation"

While the exact function of the PBBS has remained elusive, a potentially useful clinical application exists for its specific ligand, PK11195, based on three observations: (1) normal brain shows only minimal binding of PK11195, (2) in CNS pathology, in vivo PK11195 binding is predominantly found on activated microglia, and (3) when labelled with carbon-11, PK11195 can be used as a ligand for positron emission tomography (PET) (Benavides et al. 1988; Junck et al. 1989; Cremer et al. 1992; Ramsay et al. 1992; Sette et al. 1993; Banati et al. 1999b).

The use of [^{11}C](*R*)-PK11195 PET to study the acute and chronic evolution of brain disease introduces the concept of "neuroinflammation" or "glial inflammation" (Graeber 2000). It is based on the consistently made observation of activated microglia in primarily non-inflammatory, neurodegenerative diseases, such as Alzheimer's disease (Cagnin et al. 2000), Parkinson's disease and others. It is experimentally well established that neuronal injury per se in the absence of any other contributing pathology, such as damage to blood vessels, evokes a rapid, highly localised activation of microglia, the brain's intrinsic macrophages, around the somata of the injured neurons and in anatomical projection areas (Kreutzberg 1996). This neuronally triggered process of microglial activation is associated with the increased expression of immune molecules, such as MHC II. Since, however, lymphocytes, infiltrating macrophages and co-stimulatory are absent, its immunological significance is likely to be distinct from that seen in classic inflammation (Graeber 2000).

Peripheral nerve transection experiments demonstrate that "neuroinflammatory" responses are projected bi-directionally along neural fibre tracts. For example, facial nerve transection leads to a retrograde neuronal reaction and rapid induction of microglial PK11195-binding sites around the somata of lesioned motoneurons facial nucleus, while after sciatic nerve transection, an anterograde response with similar time course occurs in the gracile nucleus in the brain stem, a projection area that contains synaptic terminals from long, ipsilaterally ascending nerve fibres (Kreutzberg 1996; Banati et al. 1997).

This principle can be observed in patients with hippocampal damage in the wake of herpes simplex encephalitis. Here, [^{11}C](*R*)-PK11195 PET shows that the distribution pattern of increased [^{11}C](*R*)-PK11195 binding closely follows projecting axonal pathways, such as the large association bundles interconnecting mesocortical areas, subicular allocortices and subcortical amygdaloid nuclei (Cagnin et al. 2001a). It demonstrates how focal damage can lead to the widespread microglial activation along almost an entire affected neuronal system, in this case limbic and associated structures. It has also been possible to predict from this glial activation pattern subsequent anatomical pattern of atrophy as shown by MR-difference imaging (Fig. 1).

These findings illustrate that the pattern of microglial activation, as a surrogate marker of neuronal damage, helps to delineate the extent to which a focal lesion impacts on distributed neural connectivities not obvious by standard structural imaging techniques. It should in future also help to understand currently unexplained differences in the cognitive deficits in patients with apparently similar brain lesions (Eslinger et

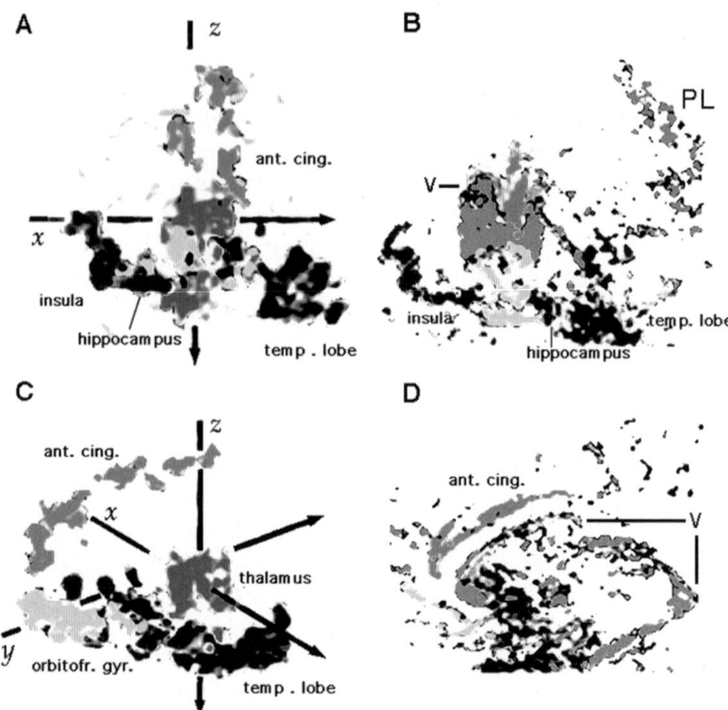

Fig. 1. A, **C** Three-dimensional, volume-rendered [^{11}C](*R*)-PK11195-binding potential map of a patient with an asymmetrical lesion of the hippocampus due to herpes simplex encephalitis. **B**, **D** Volume-rendered MRI subtraction image of the same patient. The MRI subtraction image delineates the areas that no longer match (due to tissue loss) when two of the patient's volumetric MR images taken 12 months apart are compared (Hajnal et al. 1995). MRI subtraction images also show areas of change that reflect shift of brain mass rather than change in volume. Here, such a change producing a low signal margin in the MRI subtraction image can be seen in the left parietal lobe (*PL*) that is not normally involved in the disease and, as [^{11}C](*R*)-PK11195-PET showed, was free of any active disease. Likewise, the loss of signal intensity around the ventricles (*v*) indicates in part a shift and in part a secondary ventricular enlargement due to the distributed atrophy elsewhere

al. 1993). There is already evidence from comparative studies of the functional deficits caused by different types of experimental lesions to the hippocampus or adjacent regions. It revealed unaccounted for damage leading to a reappraisal of the long-assumed role of the hippocampus in various aspects of memory (Goulet et al. 1998; Murray et al. 1998). A closer investigation of the distribution pattern of activated microglia should further improve the histopathological description of the effects of neurotoxic lesions as compared to focal structural lesions with respect to the observed variations in the functional outcome.

10.4 What Is the Significance of a Persistent "Neuroinflammatory" Response?

Our studies with [^{11}C](*R*)-PK11195 PET in herpes simplex encephalitis patients have revealed that [^{11}C](*R*)-PK11195 signals can remain elevated for many months if not years and continue to spread along the initially affected neural pathways (Cagnin et al. 2001a). However, a persistent inflammatory tissue response does not necessarily imply continued viral activity; viral products tend to disappear after the first 3 weeks and viral genome can often not be detected consistently (Sobel et al. 1986; Esiri et al. 1995). This might suggest that late [^{11}C](*R*)-PK11195 signals are primarily a delayed "neuroinflammatory" response to Wallerian-type degeneration of axons and synaptic terminals of the initially damaged neurons. Another experimental observation in mice is that neuronal death alone without blood–brain barrier damage can lead to the recruitment of peripheral T-lymphocytes and lead to the appearance of a classical inflammatory reaction despite the absence of any infectious agents (Raivich et al. 1998).

Some neuropathological evidence also comes from patients with 1-methyl-4-phenyl-1,2,3,6-tetrahydropyridine (MPTP)-induced parkinsonism, suggesting that even years after a single toxic event, activated microglia are present in selective areas indicating continuing or secondary neurodegenerative processes (Langston et al. 1999). This raises the issue to what extent the apparent spread of tissue pathology beyond the primarily affected neural pathways represents merely the delayed, full emergence of the initially sustained damage or, instead, gives evidence of more dynamic, trans-synaptic "knock-on" effects on other not directly injured neural networks, as have been reported for lesions in the visual system and in experimental lesion models of Huntington's disease (Topper et al. 1993; Cowey et al. 1999). Functionally important, late trans-synaptic microstructural changes are also found in the thalamus of limb-amputated primates where they may occur without significant neuronal cell death and are thought to underlie at least in part the cortical plasticity induced by the injury (Jones 2000). Hypothetically, the persistence of activated glia may suggest that even in the chronic stages of the recovery process following brain damage, structural plasticity continues to occur, theoretically allowing for functional plasticity even in the post-acute stages of disease (Knecht et al. 1998; Schallert et al. 2000).

10.5 Potential Role of Microglia and the "Peripheral Benzodiazepine Binding Site" in Neuronal Regeneration

Following activation, microglia change their morphology, proliferate, synthesize numerous bioactive proteins, such as cytokines, and may release neurotoxic metabolites (Banati et al. 1993; Giulian and Corpuz 1993). The latter observation, in particular, has led to the hypothesis that the activation of microglia is an important pathway in inflammatory as well as neurodegenerative brain diseases, which may result in the loss of neurons through prolonged microglia-mediated damage leading to further progression of disease (Banati et al. 1994; McGeer and McGeer 1995). More recently, activated microglia have also been attributed a role in regenerative tissue remodelling (Banati and Graeber 1994). In the context of the present study, it is an intriguing observation that CNS injury causes an increase in the synthesis of neurosteroids with a time course similar to that of the injury-induced up-regulation of the PBBS, i.e. the sites to which PK11195 binds. PBBS are known to regulate the synthesis of neurosteroids and through this pathway may influence neuronal functioning and a possible direct participation in CNS regeneration has been suggested (Lacor et al. 1999; di Michele et al. 2000).

Given these important roles in neuronal re- and degeneration, in vivo-imaging of activated microglia may potentially be a useful measure of the scope of possible functional rehabilitation and therapeutic efficacy in patients with brain injury. The particular value of $[^{11}C](R)$-

PK11195 PET lies in its ability to bind to brain macrophages/microglia as sensors of neuronal injury. In addition to structural neuro-imaging, it provides a cell biological measure of active disease, the resulting neuronal disconnection, and the protracted large-scale re-organisation of the brain following damage to specific regions or pathways. Its application in a variety of brain diseases, such as stroke (Pappata et al 2000), multiple sclerosis (Banati et al. 2000), autoimmune encephalitis (Banati et al. 2000), vasculitis (Goerres et al. 2001), ageing (Cagnin et al. 2001b), Alzheimer's disease (Cagnin et al. 2000), and typical and atypical Parkinson's disease (Banati et al. 1999a) demonstrates the generic value of imaging cellular pathology in vivo. Future work will involve the development of improved new ligands suitable for an "in vivo neuropathology", appropriate biomathematical modelling and signal quantification (Turkheimer et al. 2000), aspects that were not covered in this outline of the basic concept.

Acknowledgements. R.B.B. was supported by the Medical Research Council, the Multiple Sclerosis Society of Great Britain and Northern Ireland, the Max Planck Institute of Neurobiology (Martinsried, Germany) and through the Deutsche Forschungsgemeinschaft grant "The mitochondrial benzodiazepine receptor as indicator of early CNS pathology, clinical application in PET". A.C. was supported by a Fellowship from the European Community (BMH4/CT98/5100) in the Training and Mobility of Researchers Programme in Biomedicine. A.G. is currently supported by an Action Research, UK, project grant.

References

Anholt RR, Pedersen PL, DeSouza EB, Snyder SH (1986) The peripheral-type benzodiazepine receptor. Localisation to the mitochondrial outer membrane. J Biol Chem 261:776–783

Banati RB, Graeber MB (1994) Surveillance, intervention and cytotoxicity: is there a protective role of microglia? Dev Neurosci 16:114–127

Banati RB, Gehrman J, Schubert P, Kreutzberg GW (1993) Cytotoxicity of microglia. Glia 7:111–118

Banati RB, Myers R, Kreutzberg GW (1997) PK ('peripheral benzodiazepine')-binding sites in the CNS indicate early and discrete brain lesions: microautoradiographic detection of [3H]PK11195binding to activated microglia. J Neurocytol 26:77–82

Banati RB, Cagnin A, Myers R, Gunn RN, Piccini P, Olanow CW, Jones T, Brooks DJ (1999a) In vivo detection of activated microglia by [11C] (*R*)-PK11195 PET indicates involvement of the globus pallidum in idiopathic Parkinson's disease. Parkinsonism and related disorders 5:S1–S122 (S56)

Banati RB, Goerres G W, Myers R, Gunn RN, Turkheimer FE, Kreutzberg GW, Brooks DJ, Jones T, Duncan JS (1999b) [^{11}C](*R*)-*PK11195*PET – imaging of activated microglia in vivo in Rasmussen's encephalitis. Neurology 53:2199–2203

Banati RB, Newcombe J, Gunn RN, Cagnin A, Turkhemer F, Heppner F, Price G, Wegner F, Giovanoni G, Miller D, Perkins D, Smith T, Hewson A, Bydder G, Kreutzberg GW, Jones T, Cuzner ML, Myers R (2000) The peripheral benzodiazepine binding site in the brain in multiple sclerosis: Quantitative in vivo-imaging of microglia as a measure of disease activity. Brain 123:2321–2337

Benavides J, Cornu P, Dennis T, Dubois A, Hauw J-J, MacKenzie ET, Sazdovitch V, Scatton B (1988) Imaging of human brain lesions with an omega 3 site radioligand. Ann Neurol 24:708–712

Cagnin A, Brooks DJ, Kennedy AM, Gunn RN, Myers R, Turkheimer FE, Jones T, Banati RB (2000) In vivo detection of activated microglia in Alzheimer's disease: a [11C] (*R*)-PK11195 PET study. Neurology [Suppl 3] 54:A476

Cagnin A, Myers R, Gunn RN, Lawrence AD, Stevens T, Kreutzberg GW, Jones T, Banati RB (2001a) In vivo-visualisation of activated glia by [11C](R)-PK11195 PET following herpes encephalitis reveals projected neuronal damage beyond the primary focal lesion. Brain 124:2014–2027

Cagnin A, Myers R, Gunn RN, Turkheimer FE, Cunningham VJ, Brooks DJ, Jones T, Banati RB (2001b) Imaging activated microglia in the ageing human brain. In: Gjedde A, Hansen SB, Knudsen GM, Paulson OB (eds) Physiological Imaging of the brain with PET. Academic Press, San Diego, pp 361–367

Chung Y-L, Williams A, Ritchie D, Changani KK, Hope J, Bell JD (1999). Conflicting MRI signals from gliosis and neuronal vacuolation in prion diseases. Neuroreport 10:3471–3477

Conway EL, Gundlach AL, Craven JA (1998) Temporal changes in glial fibrillary acidic protein messenger RNA and [3H]PK11195 binding in relation to imidazoline-I2-receptor and alpha 2-adrenoreceptor binding in the hippocampus following transient global forebrain ischaemia in the rat. Neuroscience 82:805–817

Cowey A, Stoerig P, Williams C (1999) Variance in transneuronal retrograde ganglion cell degeneration in monkeys after removal of striate cortex: effect of size of the cortical lesion. Vision Res 39:3642–3652

Cremer JE, Hume SP, Cullen BM, Myers R, Manjil LG, Turton DR, Luthra SK, Bateman DM, Pike VW (1992) The distribution of radioactivity in brains of rats given [N-methyl-11C]PK 11195 in vivo after induction of a cortical ischaemic lesion. Int J Rad Appl Instrum B 19:159–166

di Michele F, Lekieffre D, Pasini A, Bernardi G, Benavides J, Romeo E (2000) Increased neurosteroids synthesis after brain and spinal cord injury in rats. Neurosci Lett 284:65–68

Dubois A, Benavides J, Peny B, Duverger D, Fage D, Goti B, et al (1988) Imaging of primary and remote ischemic and excitotoxic brain lesions. An autoradiographic study of peripheral type benzodiazepine binding sites in the rat and cat. Brain Res 445:77–90

Esiri MM, Drummond CW, Morris CS (1995) Macrophages and microglia in HSV 1 infected mouse brain. J Neuroimmunol 62:201–205

Eslinger PJ, Damasio H, Damasio AR, Butters N (1993) Non verbal amnesia and asymmetric cerebral lesions following encephalitis. Brain Cogn 21:140–152

Gavish M, Bachman I, Shoukrun R, Katz Y, Veenman L, Weisinger G, Weizman A (1999) Enigma of the peripheral benzodiazepine receptor. Pharmacol Rev 51: 629–650

Giulian D, Corpuz M (1993) Microglial secretion products and their impact on the nervous system. Adv Neurol 59:315–320

Goerres GW, Revesz T, Duncan J, Banati RB (2001) Imaging cerebral vasculitis in refractory epilepsy using [11C] (*R*)-PK11195 PET. Am J Roentgenol 176:1016–1018

Goulet S, Dore FY, Murray EA (1998) Aspiration lesions of the amygdala disrupt the rhinal corticothalamic projection system in rhesus monkeys. Exp Brain Res 119:131–140

Graeber BM (2000) Glial inflammation in neurodegenerative diseases. Immunology 101[Suppl 1]:52

Hajnal JV, Saeed N, Oatridge A, Williams EJ, Young IR, Bydder GM (1995) Detection of subtle brain changes using subvoxel registration and subtraction of serial MR images. J Comput Assist Tomogr 19:677–691

Hardwick M, Fertikh D, Culty M, Li H, Vidic B, Papadopoulos V (1999) Peripheral-type benzodiazepine receptor (PBR) in human breast tissue: correlation of breast cancer cell aggressive phenotype with PBR expression, nuclear localization and PBR-mediated cell proliferation and nuclear transport of cholesterol. Cancer Res 59:831–842

Hertz L (1993) Binding characteristics of the receptor and coupling to transport proteins. In: Giessen-Crouse E (ed) Peripheral benzodiazepine receptors. Academic Press, London, pp 27–51

Itzhak Y, Baker L, Norenberg (1996) Characterization of the peripheral-type benzodiazepine receptors in cultured astrocytes labeled with [3H]PK 11195: evidence for multiplicity. Glia 9:211–218

Jones EG (2000) Cortical and subcortical contributions to activity-dependent plasticity in primate somatosensory cortex. Annu Rev Neurosci 23:1–37

Junck L, Olson JM, Ciliax BJ, Koeppe RA, Watkins GL, Jewett DM, McKeever PE, Wieland DM, Kilbourn MR, Starosta Rubinstein S, et al (1989) PET imaging of human gliomas with ligands for the peripheral benzodiazepine binding site. Ann Neurol 26:752–758

Knecht S, Henningsen H, Hohling C, Elbert T, Flor H, Pantev C, Taub E (1998) Plasticity of plasticity? Changes in the pattern of perceptual correlates of reorganization after amputation. Brain 121:717–724

Kreutzberg GW (1996). Microglia: a sensor for pathological events in the CNS. TINS 19:312–318

Krueger KE, Papadopoulos A (1993) The role of mitochondrial benzodiazepine receptors in steroidogenesis. In: Giessen-Crouse E (ed) Peripheral benzodiazepine receptors. Academic Press, London, pp 87–111

Kuhlmann AC, Guilarte TR (2000) Cellular and subcellular localization of peripheral benzodiazepine receptors after trimethyltin Neurotoxicity. J Neurochem 4:1694–1704

Lacor P, Gandolfo P, Tonon MC, Brault E, Dalibert I, Schumacher M, Benavides J, Ferzaz B (1999) Regulation of the expression of peripheral benzodiazepine receptors and their endogenous ligands during rat sciatic nerve degeneration and regeneration: a role for PBR in neurosteroidogenesis. Brain Res 815:70–80

Langston JW, Forno LS, Tetrud J, Reeves AG, Kaplan JA, Karluk D (1999) Evidence of active nerve cell degeneration in the substantia nigra of humans years after 1-methyl-4-phenyl-1,2,3,6-tetrahydropyridine exposure. Ann Neurol 46:598–605

McGeer PL, McGeer EG (1995) The inflammatory response system of brain: implications for therapy of Alzheimer and other neurodegenerative diseases. Brain Res Brain Res Rev 21:195–218

Murray EA, Barker MG, Gaffan D (1998) Monkeys with rhinal cortex damage or neurotoxic hippocampal lesions are impaired on spatial scene learning and object reversals. Behav Neurosci 112:1291–1303

Myers R, Manjil LG, Cullen BM, Price GW, Frackowiak RSJ, Cremer JE (1991) Macrophage and astrocyte populations in relation to [^{3}H]PK 11195 binding in rat brain cortex following a local ischaemic lesion. J Cereb Blood Flow Metab 11: 314–332

Olson JM, Ciliax BJ, Mancini WR, Young AB (1988) Presence of peripheral-type benzodiazepine binding sites on human erythrocyte membranes. Eur J Pharmacol 1522:47–53

Pappata S, Levasseur M, Gunn RN, Myers R, Crouzel C, Syrota A, Jones T, Kreutzberg GW, Banati RB (2000) Thalamic microglial activation in ischemic stroke detected in vivo by PET and [11C] PK11195. Neurology 55:1052–1054

Park CH, Carboni E, Wood PL, Gee KW (1996) Characterization of peripheral benzodiazepine type sites in a cultured murine BV-2 microglial cell line. Glia 16:65–70

Pawlikowski M (1993) Immunomodulatory effects of peripherally acting benzodiazepines. In: Giessen-Crouse E (ed) Peripheral benzodiazepine receptors. Academic Press, London, pp 125–133

Raivich G, Jones LL, Kloss CUA, Werner A, Neumann H, Kreutzberg GW (1998) Immune surveillance in the injured nervous system: T-lymphocytes invade the axotomized mouse facial motor nucleus and aggregate around sites of neuronal degeneration. J Neurosci 18:5804–5816

Ramsay SC, Weiller C, Myers R, Cremer JE, Luthra SK, Lammertsma AA, Frackowiak RSJ (1992) Monitoring by PET of macrophage accumulation in brain after ischaemic stroke. Lancet 339:1054–1055

Schallert T, Leasure JL, Kolb B (2000) Experience-associated structural events, subependymal cellular proliferative activity, and functional recovery after injury to the central nervous system. J Cereb Blood Flow Metab 20:1513–1528

Sette G, Baron JC, Young AR, Miyazawa H, Tillet I, Barre L, Travere JM, Derlon JM, MacKenzie ET (1993) In vivo mapping of brain benzodiazepine receptor changes by positron emission tomography after focal ischemia in the anesthetized baboon. Stroke 24:2046–2057

Shah F, Pike VW, Ashworth S, McDermott J (1994) Synthesis of the enantiomer of [N-methyl-^{11}C]PK11195 and comparison of their behaviours as PK (peripheral benzodiazepine) binding site radioligands in rats. Nucl Med Biol 21:573–581

Sobel RA, Collins AB, Colvin RB, Bhan AK (1986) The in situ cellular immune response in acute herpes simplex encephalitis. Am J Pathol 125:332–338

Stephenson DT, Schober DA, Smalstig EB, Mincy RC, Gehlert DR, Clemens JA (1995) Peripheral benzodiazepine receptors are colocalized with activated microglia following transient global forebrain ischemia in the rat. J Neurosci 15:5263–5274

Topper R, Gehrmann J, Schwarz M, Block F, Noth J, Kreutzberg GW (1993) Remote microglial activation in the quinolinic acid model of Huntington's disease. Exp Neurol 123:271–283

Turkheimer FE, Banati RB, Visvikis D, Aston JAD, Gunn RN, Cunningham VJ (2000) Modeling Dynamic PET-SPECT studies in the wavelet domain. J Cereb Blood Flow Metab 20:879–893

11 Chemokine Receptors on Mononuclear Phagocytes in the Central Nervous System of Patients with Multiple Sclerosis

C. Trebst, T.L. Sørensen, P. Kivisäkk, M.K. Cathcart,
J. Hesselgesser, R. Horuk, F. Sellebjerg, H. Lassmann,
R.M. Ransohoff

11.1 Introduction 193
11.2 Materials and Methods 195
11.3 Results 199
11.4 Discussion 204
References 208

y11.1 Introduction

Accumulation and activation of mononuclear phagocytes in the human central nervous system (CNS) is thought to be a crucial step in the pathological cascade of multiple sclerosis (MS), which frequently culminates in irreversible injury to myelin and axons (Sørensen and Ransohoff 1998). Although MS pathology is heterogeneous among patients (Lucchinetti et al. 2000), it still remains clear that destruction of myelin and axons as well as oligodendrocyte cell-death are directly related to numbers of activated inflammatory cells (Ferguson et al. 1997; Trapp et al. 1998; Bitsch et al. 1999, 2000). Therefore, the determinants of

C. Trebst and T.L. Sørensen contributed equally to this work.

monocyte recruitment to the CNS in MS and similar pathologies have been examined. Chemokines and their receptors have been implicated in monocyte trafficking under pathological as well as physiological conditions and have emerged as salient targets for investigation (Luster 1998; Zlotnik and Yoshie 2000).

The possible role of CC chemokine receptor 1 (CCR1), CCR5 and their ligands in the pathogenesis of MS was first suggested by observations in experimental autoimmune encephalomyelitis (EAE), an animal model for MS. Inhibition/blockade of macrophage inflammatory protein (MIP)-1α (CCL3), a ligand for CCR1 and CCR5, prevented the development of both acute and relapsing paralytic symptoms and infiltration of mononuclear cells into the CNS. Importantly, anti-MIP-1α did not affect the activation of encephalitogenic T cells, suggesting specificity of MIP-1α for chemoattraction of mononuclear inflammatory cells into the CNS in EAE mice (Karpus et al. 1995; Karpus and Kennedy 1997; Kennedy et al. 1998). Mice lacking CCR1 (CCR1$^{-/-}$) developed significantly reduced incidence and severity of EAE when compared with wild type littermates. CCR1$^{-/-}$ spinal cords exhibited less-dense cellular infiltrates than cords from symptomatic wild-type mice (Rottman et al. 2000). A recent report indicated that a small-molecule antagonist of human CCR1 reduced the severity of EAE in rats, although its efficacy was limited by the relative species-specificity of the reagent (Hesselgesser and Horuk 2001).

In contrast, CCR5 appears dispensable for the development of EAE, since CCR5-deficient mice are susceptible to EAE. Further, individuals homozygous for a non-functional D32 allele of CCR5 develop MS (Bennetts et al. 1997). CCR5 may, however, have a role in determining MS severity, as distinguished from MS susceptibility: genetic studies showed that individuals heterozygous for the D32 CCR5 allele experienced prolonged disease-free intervals, compared to individuals with a fully functional CCR5 receptor (Barcellos et al. 2000; Sellebjerg et al. 2000). Consequently, both CCR1 and CCR5 may be implicated in MS pathogenesis, but the relationship between each receptor and disease susceptibility and/or severity may be complex.

The hematogenous inflammatory component in MS can be examined and characterized in tissue sections as perivascular and parenchymal inflammatory cells. CNS-infiltrating leukocytes can also be identified in the CSF. Therefore, we approached our investigation (Trebst et al. 2001) of CCR1 and CCR5 in MS in two ways: (1) CCR1 and CCR5 expression on circulating and CSF CD14$^+$ monocytes was examined by flow cytometry; (2) Quantitative immunohistochemistry was applied to characterize chemokine receptor-positive cells in MS tissue sections during lesion evolution. The results of these studies implicated CCR1$^+$/CCR5$^+$ cells as infiltrating and activated mononuclear phagocytes in MS.

11.2 Materials and Methods

11.2.1 Flow Cytometry

Flow cytometry studies evaluating CCR1 and CCR5 expression on circulating and CSF monocytes, or co-expression of CCR1 with CCR5, were performed in 24 patients with monosymptomatic optic neuritis (ON) and 26 patients with MS. In addition, 24 patients with other non-inflammatory neurological diseases who underwent diagnostic lumbar puncture (LP) were included as controls (CON). ON patients had no history of neurological symptoms and were diagnosed using established clinical criteria. MS diagnosis was based on published criteria for clinical research (Poser et al. 1983). The patients underwent LP and phlebotomy at the Glostrup Hospital, Denmark or the Department of Neurology, Cleveland Clinic Foundation. Patient characteristics are summarized in Table 1. The Scientific Ethics Committee of the Government of Denmark and the Cleveland Clinic Institutional Review Board approved this study and informed consent was obtained from all participants.

Table 1. Flow cytometry studies: patient demographics

Diagnosis	Gender	Age (years)	CSF pleocytosis	CSF oligoclonal bands
ON (24)	14 F/10 M	36 (22–54)	16 (67%)	17 (71%)
MS (26)	20 F/6 M	41 (18–61)	18 (69%)	24 (92%)
CON (24)	16 F/8 M	54 (23–79)	3 (13%)	0 (0%)[a]

CON, neurological controls; F, female; M, male; MS, multiple sclerosis; ON, optic neuritis.

[a]Oligoclonal bands were not tested in one control patient.

11.2.2 Preparation of Cells, Staining and Flow Cytometry

Preparation of peripheral blood mononuclear or CSF mononuclear cells for flow cytometry was performed as previously described (Sørensen et al. 1999).

The following antibodies were used: Phycoerythrin (PE)-conjugated anti-CCR1 (Clone 53504.111, R&D Systems, Minneapolis, Minn., USA), Fluorescein isothiocyanate (FITC)- and PE-conjugated anti-CCR5 (Clone 2D7, BD PharMingen, San Diego, Calif., USA), Allophycocyanin (APC) and Peridinin chlorophyll protein (PerCP)-conjugated anti-CD14 (clone MøP9, BD Biosciences), fluorescein isothiocyanate (FITC)-conjugated anti-CD3 (clone SK7, BD Biosciences) and PE and FITC-conjugated mouse isotype controls (BD Biosciences).

11.2.3 MS Autopsy Material

We focused these studies on active MS lesions, since cellular infiltration is proposed to be an initiating event in acute lesions. These studies, therefore, afforded an opportunity to examine the fate of CCR1^{+}/CCR5^{+} monocytes that had newly entered the CNS. Paraffin-embedded archival autopsy material of five MS patients was available. The material was collected and characterized by detailed neuropathological examination at the Brain Research Institute, University of Vienna, Austria. All five cases showed a prominent deposition of immunoglobulins and complement C9neo antigen at sites of active myelin destruction. Together with abundant T-cell and macrophage infiltrates and active demyelination, this form of tissue injury has been designated pattern II (Lucchinetti et al. 2000).

In a total of 10 tissue sections, 23 active lesions were identified and according to previously published criteria the following stages of demyelinating activity were defined (Brück et al. 1994, 1995; Lucchinetti et al. 1996):

Early-active (EA) regions were located at borders between demyelinating plaques and periplaque white matter.

In late-active (LA) areas, myelin degradation was more advanced. Inactive (IA) areas showed complete demyelination. Macrophages con-

Table 2. Autopsy material: patient characteristics

Case	Sex	Age[a]	Disease course	Duration[b]	Tissue sections	Lesions	Early-active zones	Late-active zones	Inactive zones	Peri-plaque WM zones
1	F	47	Acute	3.5	2	3	3	2	2	2
2	F	46	Acute	0.4	3	4	4	4	4	3
3	F	28	SP	12	2	7	6	4	3	2
4	M	52	Acute	1.5	2	7	6	3	2	2
5	F	34	SP	144	1	2	2	2	2	1
	4 F/1 M				10	23	21	15	13	10

F, female; M, male; SP, secondary progressive; WM, white matter.
[a]Age is given in years.
[b]Disease duration is given in months.

tained either empty vacuoles or PAS (periodic acid-Schiff reaction)-positive degradation products.

In 11 of 23 lesions, all three stages of demyelinating activity could be identified. Four lesions contained EA and LA areas. Six lesions were entirely EA and 2 lesions IA. Ten representative regions outside lesions, showing neither macroscopic nor histological evidence of demyelination, were identified as internal controls (periplaque white matter) (Table 2).

11.2.4 Immunocytochemical Techniques

Immunocytochemical analysis was performed using an avidin-biotin-horseradish peroxidase complex procedure and 3,3-diaminobenzidine (DAB) as described previously (Sørensen et al. 1999). Rabbit polyclonal anti-CCR1 antibodies were generated and characterized at Berlex Biosciences (Hesselgesser et al. 1997). Murine monoclonal anti-human CCR5 (Clone 45549.111, mouse IgG_{2B}) was obtained from R&D Systems, murine monoclonal anti-human MRP14 from Bachem Bioscience Inc., King of Prussia, Pa., USA (Clone S 36.48, mouse IgG_1) and murine monoclonal anti-CD68 (Clone KP1, mouse IgG_1) from DAKO Corporation, Carpinteria, Calif., USA. Primary antibodies were omitted in controls.

For colocalization studies (CCR1 with CCR5, CCR1 with MRP14 and CCR5 with CD68), sections were simultaneously labeled with primary antibodies and then incubated with Texas red- and FITC-conjugated secondary antibodies (Southern Biotechnology Associates, Inc., Birmingham, Ala., USA). In controls, primary antibodies were omitted, and tests for cross-reactivity by secondary antibodies were performed. Sections were analyzed on a Leica TCS-NT confocal scanning laser microscope or a Leica DMR microscope (Leica Wetzlar, Heidelberg, Germany).

11.2.5 Morphometric Analysis

The number of immunostained cells was determined in at least four standardized fields (146,200 μm^2, defined by a morphometric grid) from each of the distinct lesional areas. Sections were photographed on a Leica DMR microscope and an Optronix Magnafire digital camera system and analyzed using Image Pro Plus (Media Cybernetics, Silver Spring, Md., USA).

11.2.6 Isolation and culture of human monocytes

In order to examine CCR1 and CCR5 expression during monocyte differentiation in vitro, monocytes were obtained from freshly donated human peripheral blood from five healthy donors [3 F/2 M, mean age: 33 (21–43) years]. The blood was immediately diluted 1:1 with phosphate-buffered solution (PBS), containing 1% human albumin (SIGMA) and underlayered with Ficoll-Paque (Pharmacia, Piscataway, N.J., USA) for separation of mononuclear cells by density centrifugation. Monocytes were separated by adherence to serum-coated flasks according to the method of Kumagai et al. and subsequently detached using 0.5 mM ethylenediamine tetra-acetic acid (EDTA) (Kumagai et al. 1979). This preparation contained more than 95% monocytes as determined by flow cytometry for CD14 surface expression. Isolated monocytes were cultured in Dulbecco's modified Eagle Medium (DMEM, Mediatech Cellgro Inc., Herndon, Va., USA) supplemented with L-Glutamine, 4.5 mg/l glucose and 10% bovine calf serum (Hyclone, Logan, Utah, USA) for 7 days at 37°C and 10% CO_2. Under these conditions, monocytes differentiate into macrophages after 7 days (Tuttle et al. 1998; Hariharan et al. 1999). Granulocyte-macrophage colony-stimulating factor (GM-CSF) was omitted in the cell culture medium since this might alter chemokine surface expression levels (Tuttle et al. 1998). After 1 day or 7 days in culture, cells were detached from the flask using 0.5 mM EDTA and gentle scraping and resuspended in PBS. CCR1 and CCR5 expression was determined by flow cytometry.

11.2.7 Statistical Analysis

Non-parametric tests (Mann-Whitney test and Wilcoxon signed rank test) were applied because the data were not normally distributed (Kolmogorov-Smirnov test). A p value less than 0.05 was considered statistically significant.

11.3 Results

11.3.1 CCR1$^+$/CCR5$^+$ Monocytes Accumulate Preferentially in the CSF Compartment

CCR1 and CCR5 expression on CD14$^+$ monocytes in peripheral blood and CSF was compared in patients with ON, MS and neurological controls (CON; see Table 1 for detailed demographics). Approximately 90% of CSF monocytes expressed CCR1 and roughly 80% expressed CCR5, which was significantly ($p<0.001$) higher compared to peripheral blood. There were no differences between the three patient groups examined.

To define the distribution of the CCR1$^+$/CCR5$^+$ phenotype on monocytes, we examined co-expression of CCR1 and CCR5 on circulating and CSF monocytes from five patients with MS and five controls (CON). Significantly more CSF monocytes (70%, 18.8%–88.7%, $p<0.01$) had the CCR1$^+$/CCR5$^+$ phenotype, compared to 19% (9.7%–34.2%) of circulating monocytes. Virtually all CCR1$^+$ CSF monocytes (96.6%) also expressed CCR5, compared to 23.3% of circulating CCR1$^+$ monocytes. By comparison, a majority of CCR5$^+$ monocytes both in the circulation and CSF were also CCR1$^+$ (79.5% and 83.9%, respectively).

Together, these results indicated that $CCR1^+/CCR5^+$ monocytes, while a minority of the circulating monocyte pool, were highly enriched in the CSF population, regardless of the presence of CNS inflammatory pathology.

11.3.2 Distribution of Mononuclear Phagocytes in Acute MS Lesions

To follow the fate of $CCR1^+/CCR5^+$ hematogenous monocytes in MS lesions, immunohistochemistry for CD68, CCR5, and CCR1 was performed on ten tissue sections of five patients with MS and distribution of immunoreactive cells in relation to demyelinating activity was established. In the 10 available tissue sections 23 lesions were identified containing 21 areas of EA demyelination, 15 LA regions and 13 IA areas (see "Material and Methods" and Table 2 for details of lesion activity determination and description of pathological material).

In EA zones, two populations of $CD68^+$ cells were observed: perivascular $CD68^+$ cells had the appearance of activated monocytes with prominent granular cytoplasm, whereas parenchymal $CD68^+$ cells exhibited predominantly the morphology of activated process-bearing microglia.

CCR5 immunoreactivity was almost exclusively found on perivascular cells and on activated monocytes in zones of early active demyelination. Perivascular $CCR5^+$ cells were comprised both of $CD68^+$ monocytes and $CD3^+$ lymphocytes. Dual-label immunofluorescence histochemistry and confocal microscopy for CD68 and CCR5 localized $CD68^+/CCR5^+$ double-positive cells to perivascular accumulations of EA regions, with CD68-single-positive microglial cells detected in the parenchyma.

In contrast to EA zones, $CD68^+$ cells in LA areas comprised a homogenous population of large, strongly CD68-immunoreactive phagocytic macrophages.

$CCR5^+$ cells in LA regions resembled large phagocytic cells and were predominantly found in the parenchyma. In serial section analysis and dual-label immunohistochemistry, $CD68^+$ phagocytic parenchymal macrophages in LA regions uniformly expressed CCR5.

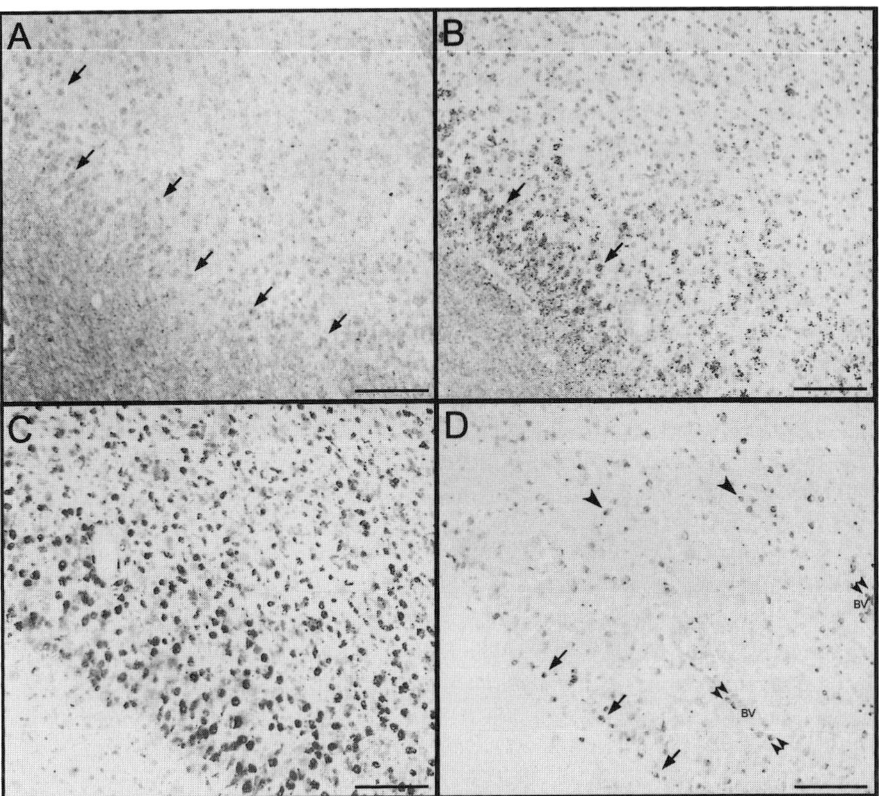

Fig. 1A–D. CCR1-expressing cells in an early-active region of a demyelinating MS lesion. Actively demyelinating and expanding lesion stained with luxol fast blue myelin stain (LFB, **A**) indicating the lesion edge and adjacent early-active area. Macrophages contained myelin degradation products including LFB (**A**, *arrows*), indicating ongoing demyelination. C9 neoantigen was readily detected on degenerating myelin sheaths (**B**, *arrows*), indicating that this was a pattern II lesion (Lucchinetti et al. 2000). CD68 immunoreactivity (**C**) revealed a population of phagocytic macrophages throughout the lesion, concentrated most densely at the lesion edge. Immunohistochemistry for CCR1 (**D**) localized many $CCR1^+$ cells to the immediate lesion edge (**D**, *arrows*). Other $CCR1^+$ cells within early-active areas were either perivascular (**D**, *double arrowheads*) or dispersed in the parenchyma (**D**, *single arrowheads*). *BV*, blood vessel; *bar*=100 μM

CD68 and CCR5 immunoreactivity in IA zones was essentially identical with that observed in LA regions.

$CCR1^+$ cells were predominantly localized at lesion edges and within areas of EA demyelination (Fig. 1). In serial section analyses, $CCR1^+$ cells colocalized with a minority subpopulation of $CD68^+$ cells, but not with CD3 immunoreactivity (Fig. 1C,D and data not shown). $CCR1^+$ cells in EA areas were either perivascular or in parenchymal foci of active demyelination (Fig. 1D). Both perivascular and parenchymal $CCR1^+$ cells exhibited the morphology of monocytes. Dual-label immunofluorescence histochemistry and confocal microscopy localized $CCR1^+/CCR5^+$ cells predominantly to perivascular cell aggregates in EA regions (data not shown). As in CSF, all $CCR1^+$ cells were $CCR5^+$, whereas $CCR5^+$ cells were not invariably $CCR1^+$. $CCR1^+$ cells at lesion edges were small, round cells that co-localized with macrophage-related protein $(MRP)14^+$ cells (data not shown).

In contrast to EA areas, only occasional cells in LA and IA regions exhibited CCR1 immunoreactivity and were predominantly perivascular (data not shown).

In summary, in EA zones mononuclear phagocytes formed a heterogeneous population of $CCR1^+/CCR5^+$ perivascular monocytes and parenchymal $CCR1^-/CCR5^-$ microglial cells. In LA and IA zones a homogenous population of monocyte- and microglia-derived mononuclear phagocytes were $CCR1^-/CCR5^+$.

11.3.3 Quantification of Mononuclear Phagocytes in Relation to Demyelinating Activity

The density of $CD68^+$ cells was significantly ($p<0.001$) higher in EA, LA and IA zones, compared with periplaque white matter, with consistent CD68 counts within the different zones of individual lesions (Table 3), indicating that the transition from EA to LA and IA demyelination in these lesions was characterized by monocyte and microglia activation and redistribution, rather than continuous accumulation of newly-infiltrating cells from the bloodstream.

EA areas contained 417±55.1 $CCR1^+$ cells/mm^2 (mean±standard error of the mean). This density was significantly ($p<0.001$), 100-fold higher than in periplaque white matter (4.9±2.2 cells/mm^2) and repre-

Table 3. Quantitation of CD68 in early-active, late-active and inactive zones of demyelination and in the periplaque white matter[a]

Case number	Tissue section	Periplaque white matter	Lesion assignment	Early-active	Late-active	Inactive
1	1A	315	Lesion 1	1,317	1,231	1,387
			Lesion 2	825		
	1B	235	Lesion 3	1,603	1,822	1,353
2	2A	318	Lesion 1	1,756	1,563	1,554
			Lesion 2	1,856	2,761	2,535
	2B	226	Lesion 3	2,704	1,866	2,708
	2C	158	Lesion 4	1,827	2,156	2,249
3	3A	158	Lesion 1	1,763	1,969	
			Lesion 2	1,260	1,587	
			Lesion 7			1,087
	3B	127	Lesion 3	2,181	2,141	2,086
			Lesion 4	1,531	1,195	1,665
			Lesion 5	2,267		
			Lesion 6	1,517		
4	4A	139	Lesion 1	1,257	1,228	
			Lesion 2	1,415	1,380	
			Lesion 3	1,247		
	4B	170	Lesion 4	1,952	1,307	1,379
			Lesion 5	1,428		
			Lesion 6	1,874		
			Lesion 7			2,266
5	5A	188	Lesion 1	1,545	1,626	1,133
			Lesion 2	1,870	1,781	1,531

mean±SEM203.3±21.823 lesions1,666.4±90.81,707.5±112.11,763.9±149.7

[a]Numbers are given in cells/mm^2; SEM, standard error of the mean.

sented a mean of 25% (5%–54%) of CD68$^+$ cells in EA areas. In LA areas, 271±54.3 cells/mm^2 (16%, 3%–44% of CD68$^+$ cells) expressed CCR1. Compared to EA and LA, IA areas had a significant (EA vs IA $p<0.001$, LA vs IA $p<0.05$) lower density of CCR1$^+$ cells (128.9±25.1 cells/mm^2; 7%, 1%–27%).

Quantitative immunohistochemical analysis of CCR5 expression in EA areas revealed 705±64 cells/mm^2 (mean±standard error). This number represented significantly ($p<0.05$) more CCR5$^+$ cells than found in periplaque white matter (65±25 cells/mm^2), and constituted a mean of 42% (range: 13%–67%) of CD68$^+$ cells within EA areas. The density of CCR5$^+$ cells was significantly ($p<0.001$) higher in areas of LA demyelination, compared with EA regions. In LA areas, a mean of 1599±145 cells/mm^2 expressed CCR5. In these regions, CCR5$^+$ cells represented 94% (62%–205%) of CD68$^+$ cells. In IA areas, CCR5 expression was 1480±181 cells/ mm^2, representing 84% (18%–152%) of CD68$^+$ cells.

11.3.4 Expression of CCR1 and CCR5 Was Differentially Regulated During Monocyte Differentiation In Vitro

In order to investigate regulation of CCR1 and CCR5 expression during monocyte maturation in vitro, chemokine receptor expression on freshly isolated monocytes and at two different time points during monocyte differentiation in culture was examined by flow cytometry. Thirty-four percent (27.9%–41%) of freshly isolated monocytes expressed CCR1, while 9.7% (4.5%–17.6%) expressed CCR5. After 7 days in culture (DIC) the cells had become enlarged or spindle shaped with extended processes. After 1 and 7 DIC, CCR1 was expressed by 4.7% (2.1%–6.7%, $p<0.01$) and 2.5% (1.5%–3.4%), respectively.

After 1 DIC, 78.7% (70.3%–87.5%) of cells expressed CCR5, as compared to 9.7% (4.5%–17.6%, $p<0.01$) of freshly isolated monocytes. After 7 DIC 63.1% (40%–83.8%) of macrophages expressed CCR5.

These in vitro results indicated that mature, adherent macrophages in vitro expressed very low levels of CCR1 and high levels of CCR5, compared to freshly isolated monocytes.

11.4 Discussion

This investigation was initiated to characterize chemokine receptor involvement in recruitment and activation of mononuclear phagocytes in the human CNS during early demyelination. The study was motivated by evidence from human material and animal models, implicating two major ligands for these receptors, RANTES/CCL5 and MIP-1α/CCL3, in the pathogenesis of inflammatory demyelination. Both chemokine receptors were highly enriched on monocytes in the CSF compartment and in perivascular cell accumulations in acute MS lesions. Furthermore, differential expression patterns for CCR1 and CCR5 were observed during lesion evolution and during culture of human monocytes in vitro. These results provide insight into pathogenetic mechanisms during early demyelinating events in MS and identify targets for future study and potential therapeutic intervention.

11.4.1 Hematogenous Monocytes That Enter the CNS Are CCR1$^+$/CCR5$^+$

In this report, we provide evidence that hematogenous monocytes that infiltrate the CNS are CCR1$^+$/CCR5$^+$. This evidence is primarily based on findings of our flow cytometry studies showing that a majority of monocytes in the CSF compartment were CCR1$^+$/CCR5$^+$. This CCR1$^+$/CCR5$^+$ population constituted a minority of the circulating pool of monocytes. Consistent with the hypothesis that the lumbar CSF is in equilibrium with the perivascular space of the CNS white matter, dual-label immunohistochemistry revealed CCR1$^+$/CCR5$^+$ perivascular cell accumulations in early demyelinating regions.

Although enrichment of CCR1$^+$/CCR5$^+$ monocytes in the CSF compartment was observed independent of CNS pathology, CCR1$^+$/CCR5$^+$ perivascular cell accumulations have not been detected in non-inflamed brain sections (unpublished observations). We therefore propose that CCR1$^+$/CCR5$^+$ monocytes will only be retained in the CNS perivascular space in the presence of appropriate ligands, such as RANTES, which is present in CSF of MS patients and is abundantly expressed by the perivascular elements of MS lesions (Sørensen et al. 1999).

CCR1$^+$ cells at MS lesion edges co-expressed the macrophage-related protein (MRP) 14, a 14-kDa calcium-binding protein expressed primarily by circulating human neutrophils and monocytes (Odink et al. 1987). In vitro, this antigen shows a decline in expression during monocyte differentiation (Goebeler et al. 1993). In MS lesions, MRP-14 expression is associated with the earliest stage of macrophage-mediated demyelinating activity (Brück et al. 1995). It is uncertain whether MRP14 can be expressed by microglia under certain pathological conditions. In MS lesions, we and others detected MRP-14 expression only on small round, non-process-bearing cells, morphologically consistent with monocytes (Brück et al. 1995). Co-expression of CCR1 and MRP-14 was therefore interpreted as identifying a population of newly recruited hematogenous monocytes.

11.4.2 Hematogenous Monocytes and Resident Microglia Show Different Chemokine Receptor Expression Patterns During Early Demyelination

In EA zones, a mean of 25% of $CD68^+$ mononuclear phagocytes expressed CCR1 while a mean of 42% of $CD68^+$ cells expressed CCR5. If the majority of hematogenous monocytes in the CNS are $CCR1^+/CCR5^+$, the remaining 58%–75% of $CCR1^-/CCR5^-/CD68^+$ mononuclear phagocytes most plausibly represent resident microglial cells. This interpretation was also supported by the distribution and morphology of $CD68^+$ cells in EA areas, where we identified two distinct populations. One $CD68^+$ population was identical to $CCR1^+/CCR5^+$ perivascular monocytes. The second population of $CD68^+$ cells exhibited the morphology of parenchymal microglia. $CD68^+$ cells with microglial morphology did not co-localize with CCR1 or CCR5 immunoreactivity and were concentrated in and around regions of EA demyelination.

Based on these data, we propose that initial effectors of demyelination in MS lesions include $CCR1^+/CCR5^+$ hematogenous monocytes as well as $CCR1^-/CCR5^-$ resident parenchymal microglial cells.

11.4.3 Lesion Evolution Is Characterized by Activation, Rather than Accumulation of Monocytes from the Blood Stream

We and others found that numbers of $CD68^+$ cells in EA, LA, and IA regions of individual active MS lesions remained virtually unchanged (Brück et al. 1995, 1996). These observations led to the interpretation that changes in the $CD68^+$ population occurred via activation and redistribution in the absence of meaningful continuous invasion of cells from the bloodstream. Another possibility is that large numbers of $CD68^+$ cells undergo apoptosis in EA regions of MS lesions and are quantitatively replaced by hematogenous cells. Arguing against this interpretation, analyses of apoptotic cells in MS lesions have not reported dying macrophages or microglial cells in numbers sufficient to offer an alternative explanation of our results (Dowling et al. 1996; Bonetti and Raine 1997).

11.4.4 Expression of CCR1 and CCR5 Are Differentially Regulated During Lesion Evolution

The number of $CCR5^+$ cells was significantly higher in LA and IA regions compared to EA regions, suggesting that CCR5 expression is not only maintained by the infiltrating hematogenous monocyte population over time but was also upregulated by resident, initially CCR5-negative, microglial cells during activation and transformation into mature macrophages. In contrast, we observed a significant lower number of CCR1-expressing monocytes in IA regions compared to EA and LA areas. How can differential regulation of CCR1 and CCR5 expression during MS lesion evolution be explained? Expression of chemokine receptors on the cell surface is a result of a complex interplay of regulatory mechanisms including cytokine stimulation, ligand-induced internalization as well as differentiation and activation of the receptor-bearing cell. Here, we report that the expression of CCR1 and CCR5 on human monocytes is differentially regulated during monocyte maturation in vitro. As reported previously, we found higher CCR5 surface expression on macrophages when compared to freshly isolated monocytes (Di Marzio et al. 1998; Tuttle et al. 1998). Our data constitute the first report that CCR1 surface expression is downregulated during monocyte differentiation in vitro. The loss of CCR1 and the gain of CCR5 expression during MS lesion evolution might therefore be explained by an increased activation and maturation of the infiltrating hematogenous monocyte population.

In considering the implications of our findings for MS pathogenesis, it is pertinent that chemokine receptors mediate effects beyond chemotaxis, some of which could be important for MS pathogenesis (Bacon et al. 1995). These effects include direct induction of IL-12 (Aliberti et al. 2000) and nitric oxide production as well as modulation of cytokine release (Fahey et al. 1992; Villalta et al. 1998; Aliberti et al. 1999).

Our results provide insight into potential mechanisms of trafficking and activation of hematogenous monocytes in the CNS of MS patients. The challenge for the future is to define the roles of these components in disease pathogenesis.

Acknowledgements. We are grateful for the support provided by the Nancy Davis Center Without Walls for a morphometric image analysis station. This study was supported by: National Institutes of Health (1PO1 NS38667 to RMR); Williams Fund for MS Research (to RMR); Deutsche Forschungsgemeinschaft, Germany (TR463/1–1 to CT); P. Carl Petersen Foundation, Denmark (to TLS); Niels Ydes Foundation, Denmark (to TLS); Bundesministerium für Bildung, Wissenschaft und Kultur, Austria (GZ 70.056/2-Pr/4/99 to HL).

References

Aliberti J, Reis e Sousa C, Schito M, Hieny S, Wells T, Huffnagle GB, Sher A (2000) CCR5 provides a signal for microbial induced production of IL-12 by CD8α^+ dendritic cells. Nat Immunol 1:83–87

Aliberti JC, Machado FS, Souto JT, Campanelli AP, Teixeira MM, Gazzinelli RT, Silva JS (1999) beta-Chemokines enhance parasite uptake and promote nitric oxide-dependent microbiostatic activity in murine inflammatory macrophages infected with Trypanosoma cruzi. Infect Immun 67:4819–4826

Bacon KB, Premack BA, Gardner P, Schall TJ (1995) Activation of dual T cell signaling pathways by the chemokine RANTES. Science 269:1727–1730

Barcellos LF, Schito AM, Rimmler JB, Vittinghoff E, Shih A, Lincoln R, Callier S, Elkins MK, Goodkin DE, Haines JL, Pericak-Vance MA, Hauser SL, Oksenberg JR (2000) CC-chemokine receptor 5 polymorphism and age of onset in familial multiple sclerosis. Multiple Sclerosis Genetics Group. Immunogenetics 51:281–288

Bennetts BH, Teutsch SM, Buhler MM, Heard RN, Stewart GJ (1997) The CCR5 deletion mutation fails to protect against multiple sclerosis. Hum Immunol 58:52–59

Bitsch A, Schuchardt J, Bunkowski S, Kuhlmann T, Brück W (2000) Acute axonal injury in multiple sclerosis. Correlation with demyelination and inflammation. Brain 123:1174–1183

Bitsch A, Wegener C, Da Costa C, Bunkowski S, Reimers CD, Prange HW, Bruck W (1999) Lesion development in Marburg's type of acute multiple sclerosis: from inflammation to demyelination. Mult Scler 5:138–146

Bonetti B, Raine CS (1997) Multiple sclerosis: oligodendrocytes display cell death-related molecules in situ but do not undergo apoptosis. Ann Neurol 42:74–84

Brück W, Porada P, Poser S, Rieckmann P, Hanefeld F, Kretzschmar HA, Lassmann H (1995) Monocyte/macrophage differentiation in early multiple sclerosis lesions. Ann Neurol 38:788–796

Brück W, Schmied M, Suchanek G, Brück Y, Breitschopf H, Poser S, Piddlesden S, Lassmann H (1994) Oligodendrocytes in the early course of multiple sclerosis. Ann Neurol 35:65–73

Brück W, Sommermeier N, Bergmann M, Zettl U, Goebel HH, Kretzschmar HA, Lassmann H (1996) Macrophages in multiple sclerosis. Immunobiology 195:588–600

Di Marzio P, Tse J, Landau NR (1998) Chemokine receptor regulation and HIV type 1 tropism in monocyte- macrophages. AIDS Res Hum Retroviruses 14:129–138

Dowling P, Shang G, Raval S, Menonna J, Cook S, Husar W (1996) Involvement of the CD95 (APO-1/Fas) receptor/ligand system in multiple sclerosis brain. J Exp Med 184:1513–1518

Fahey TJ, III, Tracey KJ, Tekamp-Olson P, Cousens LS, Jones WG, Shires GT, Cerami A, Sherry B (1992) Macrophage inflammatory protein 1 modulates macrophage function. J Immunol 148:2764–2769

Ferguson B, Matyszak MK, Esiri MM, Perry VH (1997) Axonal damage in acute multiple sclerosis lesions. Brain 120:393–399

Goebeler M, Roth J, Henseleit U, Sunderkotter C, Sorg C (1993) Expression and complex assembly of calcium-binding proteins MRP8 and MRP14 during differentiation of murine myelomonocytic cells. J Leukoc Biol 53:11–18

Hariharan D, Douglas SD, Lee B, Lai JP, Campbell DE, Ho WZ (1999) Interferon-gamma upregulates CCR5 expression in cord and adult blood mononuclear phagocytes. Blood 93:1137–1144

Hesselgesser J, Halks-Miller M, DelVecchio V, Peiper SC, Hoxie J, Kolson DL, Taub D, Horuk R (1997) CD4-independent association between HIV-1 gp120 and CXCR4: functional chemokine receptors are expressed in human neurons. Curr Biol 7:112–121

Hesselgesser J, Horuk R (2001) The CCR1 antagonist BX 471 is effective in animal models of multiple sclerosis and organ transplant rejection. Keystone symposium: Chemokines and chemokine receptors. Abstract 20

Karpus WJ, Kennedy KJ (1997) MIP-1alpha and MCP-1 differentially regulate acute and relapsing autoimmune encephalomyelitis as well as Th1/Th2 lymphocyte differentiation. J Leukoc Biol 62:681–687

Karpus WJ, Lukacs NW, McRae BL, Strieter RM, Kunkel SL, Miller SD (1995) An important role for the chemokine macrophage inflammatory protein-1 alpha in the pathogenesis of the T cell-mediated autoimmune disease, experimental autoimmune encephalomyelitis. J Immunol 155:5003–5010

Kennedy KJ, Strieter RM, Kunkel SL, Lukacs NW, Karpus WJ (1998) Acute and relapsing experimental autoimmune encephalomyelitis are regulated by differential expression of the CC chemokines macrophage inflammatory protein-1alpha and monocyte chemotactic protein-1. J Neuroimmunol 92:98–108

Kumagai K, Itoh K, Hinuma S, Tada M (1979) Pretreatment of plastic Petri dishes with fetal calf serum. A simple method for macrophage isolation. J Immunol Methods 29:17–25

Lucchinetti C, Brück W, Parisi J, Scheithauer B, Rodriguez M, Lassmann H (2000) Heterogeneity of multiple sclerosis lesions: implications for the pathogenesis of demyelination. Ann Neurol 47:707–717

Lucchinetti CF, Brück W, Rodriguez M, Lassmann H (1996) Distinct patterns of multiple sclerosis pathology indicates heterogeneity on pathogenesis. Brain Pathol 6:259–274

Luster AD (1998) Chemokines – chemotactic cytokines that mediate inflammation. N Engl J Med 338:436–445

Odink K, Cerletti N, Bruggen J, Clerc RG, Tarcsay L, Zwadlo G, Gerhards G, Schlegel R, Sorg C (1987) Two calcium-binding proteins in infiltrate macrophages of rheumatoid arthritis. Nature 330:80–82

Poser CM, Paty DW, Scheinberg L, McDonald WI, Davis FA, Ebers GC, Johnson KP, Sibley WA, Silberberg DH, Tourtellotte WW (1983) New diagnostic criteria for multiple sclerosis: guidelines for research protocols. Ann Neurol 13:227–231

Rottman JB, Slavin AJ, Silva R, Weiner HL, Gerard CG, Hancock WW (2000) Leukocyte recruitment during onset of experimental allergic encephalomyelitis is CCR1 dependent. Eur J Immunol 30:2372–2377

Sellebjerg F, Madsen HO, Jensen CV, Jensen J, Garred P (2000) CCR5 delta32, matrix metalloproteinase-9 and disease activity in multiple sclerosis. J Neuroimmunol 102:98–106

Sørensen TL, Ransohoff RM (1998) Etiology and pathogenesis of multiple sclerosis. Semin Neurol 18:287–294

Sørensen TL, Tani M, Jensen J, Pierce V, Lucchinetti C, Folcik VA, Qin S, Rottman J, Sellebjerg F, Strieter RM, Frederiksen JL, Ransohoff RM (1999) Expression of specific chemokines and chemokine receptors in the central nervous system of multiple sclerosis patients. J Clin Invest 103:807–815

Trapp BD, Peterson J, Ransohoff RM, Rudick R, Mörk S, Bö L (1998) Axonal transection in the lesions of multiple sclerosis. N Engl J Med 338:278–285

Trebst C, Sørensen T, Kivisäkk P, Cathcart M, Hesselgesser J, Horuk R, Sellebjerg F, Lassmann H, Ransohoff RM (2001) $CCR1^+/CCR5^+$ mononuclear phagocytes accumulate in the central nervous system of patients with multiple sclerosis. Am J Path 159:1701–1710

Tuttle DL, Harrison JK, Anders C, Sleasman JW, Goodenow MM (1998) Expression of CCR5 increases during monocyte differentiation and directly mediates macrophage susceptibility to infection by human immunodeficiency virus type 1. J Virol 72:4962–4969

Villalta F, Zhang Y, Bibb KE, Kappes JC, Lima MF (1998) The cysteine-cysteine family of chemokines RANTES, MIP-1alpha, and MIP- 1beta induce trypanocidal activity in human macrophages via nitric oxide. Infect Immun 66:4690–4695

Zlotnik A, Yoshie O (2000) Chemokines: a new classification system and their role in immunity. Immunity 12:121–127

12 The Role of Apoptosis in Neuroinflammation

F. Zipp, O. Aktas, J.D. Lünemann

12.1 Neuroinflammatory Diseases 213
12.2 Dual Role of Apoptosis in Neuroinflammation 215
12.3 The Role of the TRAIL System in Neuroinflammation 219
12.4 The Role of Apoptosis in Established Therapies in MS 221
References 222

12.1 Neuroinflammatory Diseases

Inflammatory diseases of the central nervous system (CNS) play a major role in clinical neurology. It is currently under debate whether inflammatory processes determine the outcome in, for example, brain injury (Woiciechowsky et al. 1998) and cerebral ischemia (Dirnagl et al. 1999). The classical and most demanding acute inflammatory disease of the CNS is bacterial meningitis which still displays a mortality rate of about 20%, despite effective antibiotic treatment (Schuchat et al. 1997). The functional outcome of bacterial meningitis regarding long-term sequelae is dictated by neuronal injury that leads to seizures, paralysis, and cognitive deficits in survivors. The most common chronic inflammatory disease of the CNS in Northern America and Europe, which causes prolonged and severe disability in young adults, is multiple sclerosis (MS). MS is thought to be an autoimmune disorder with demyelination and axonal pathology leading to clinical symptoms (Noseworthy et al. 2000). The pathogenesis in both bacterial meningitis and multiple sclerosis has not been completely elucidated, and therapies are still inefficient despite the progress made thus far. The heterogeneous clinical course of both meningitis and multiple sclerosis requires individual treatment. Since overall therapeutic options are still lacking, clinical and experimental approaches have been aimed towards identifying genes to advance pathogenetic explanation, discover predisposing parameters for various forms of the disease, and elicit suitable therapeutic strategies.

Bacterial meningitis is a relatively rare but severe disease affecting mostly young adults and children (2 out of 100,000 in Europe and the USA and 10 out of 100,000 in developing countries). While the onset of the disease is caused by bacteria and their components, *Streptococcus pneumoniae* being the most common pathogen of meningitis, elimination of the infectious agents leads to an immune response which shares features of autoimmune CNS disease. Neuronal damage responsible for disability has been observed in the dentate gyrus (Zysk et al. 1996) and the cortex (Leib et al. 1996). In pneumococcal meningitis, the direct toxic effects of bacteria (e.g., pneumolysin) and the host inflammatory response are critical for the overall resulting damage (Schumann et al. 1998; Freyer et al. 1999). Although there is a significant body of evidence that leukocytes play a key role in brain injury due to bacterial meningitis (Tuomanen et al. 1989), little is known about leukocyte-dependent mediators causing or contributing to this tissue damage (Braun and Tuomanen 1999). Blocking leukocyte entry into the cerebrospinal fluid (CSF) inhibits about 50% of the neuronal injury (Braun et al. 1999) and strongly suggests the role of soluble leukocyte-derived factors. Based on the fact that the improvement of antibiotics and intensive care has had no effect on the outcome of the disease over the last 40 years, the pathophysiological cascade evoked by bacteria, their components, and leukocyte-dependent mediators are the focus of current research activities.

MS has so far been considered an autoimmune demyelinating disease of the CNS (Martin et al. 1992; Martino and Hartung 1999). Yet, it has very recently become clear that axonal/neuronal damage is also involved in the pathology (Ferguson et al. 1997; Kornek et al. 2000; Smith et al. 2000). T cells specific for myelin antigens are thought to initiate (Hohlfeld et al. 1995) and, in cooperation with B cells, perpetuate the immune processes of this disease (Noseworthy et al. 2000). Our current multi-step model for the initiation of T cell-mediated autoimmune inflammatory disease of the CNS includes peripheral activation of T cells specific for myelin antigens and T helper (Th) 1-type differentiation (Karpus and Ransohoff 1998). Once activated, autoreactive T cells will cross the blood–brain barrier and respond to CNS antigens in situ. The process of lesion formation in the CNS is further governed by a complex expression pattern of adhesion molecules and their ligands, cyto- and chemokines as well as enzymes by leukocytes, cerebrovascular endothelium, and parenchymal cells of the CNS (Raine and Canella 1992; Karpus and Ransohoff 1998). Experimental autoimmune encephalomyelitis (EAE) serves as a common animal model for autoimmune-mediated inflammatory CNS disease, sharing some features with MS in humans (Wekerle et al. 1994). Although many aspects of the proinflammatory events of EAE have been elucidated, the exact mechanisms underlying tissue damage and disease remission is not fully understood.

One approach for the treatment of autoimmune disorders is augmenting tolerance by immunoregulation. Strategies in this direction include induction of specific immunological tolerance via anergizing and deletion or suppression of autoreactive clones (van Parijs et al. 1998; Shevach 2000). A mechanism of tolerance that has received much attention recently due to the discovery of new death receptors and ligands is apoptosis. The major way of T-cell deletion including elimination of disease-mediating Th1 cells is activation-induced cell death (AICD), an apoptotic form of cell death.

12.2 Dual Role of Apoptosis in Neuroinflammation

In neuroinflammation, apoptosis takes on two roles, determining T-cell fate on the one hand and mediating tissue damage on the other (Zipp et al. 1999). An important mechanism by which the immune system controls autoreactive T cells, which upon activation are potentially harmful, is AICD (van Parijs et al. 1998). A more general function of AICD is the limitation of immune responses to avoid excessive clonal expansion (Lenardo et al. 1999). The initiation of cell cycling in response to antigen mediated by interleukin (IL)-2 signaling renders T cells susceptible to AICD. During T-cell activation, death ligands are upregulated which induce T-cell death in a suicidal or fratricidal manner. For AICD induction, the key death ligand is CD95 (APO-1/Fas) ligand (Dhein et al. 1995; Brunner et al. 1995; Krammer 1998; Krammer et al. 1998), but a role for tumor necrosis factor (TNF)-α was also reported (Sytwu et al. 1996).

Moreover, T cells are controlled by an active immune privilege in the CNS, which normally wards off invading T cells. One mechanism is that these T cells undergo apoptosis after invasion into the CNS (Pender et al. 1991; Schmied et al. 1993) and are thus eliminated via death receptor/ligand interactions, e.g., TNF receptor/TNF (Bachmann et al. 1999) or CD95/CD95 ligand (Bechmann et al. 1999, 2000; Gold et al. 1997; Flügel et al. 2000). Whether other death ligands, such as the TNF-related apoptosis inducing ligand (TRAIL/APO-2 ligand), are involved in the maintenance of CNS immune privilege is a matter that is currently being investigated.

The second role of apoptosis is its involvement in the inflammatory damage mechanisms in the CNS. Both oligodendrocytes and neurons become susceptible to apoptotic cell death induced by death ligands in an inflammatory context. Human oligodendrocytes have been observed to exhibit the capacity of undergoing CD95-mediated apoptosis upon IFN-γ treatment (Pouly et al. 2000). The oligodendrocyte apoptosis could be enhanced by TNF-α and blocked by a caspase inhibitor. In MS, elevated CD95 expression on oligodendrocytes was reported in a post-mortem study (D'Souza et al. 1996). In EAE, a transgenic approach inhibiting caspase activation selectively in oligodendrocytes protected mice against the disease (Hisahara et al. 2000). The inflammatory demyelination seen in EAE has recently been demonstrated to be accompanied by acute apoptotic neuronal cell loss (Meyer et al. 2001). Medana et al. showed that neurons were capable of undergoing apoptosis induced by cytotoxic $CD8^+$ T cells when the neurons were rendered antigen-presenting cells upon incubation with the relevant antigen peptide (Medana et al. 2000). This T-cell effector apoptosis was dependent on CD95/CD95 ligand interaction.

Data from an animal model of bacterial meningitis showed neuronal apoptosis in the hippocampus (Braun et al. 1999). Preventive caspase inhibition markedly reduced the destruction in the brain and confirmed the involvement of apoptotic death of CNS cells in the acute neuroinflammation.

The contribution of apoptotic mechanisms to T-cell regulation and to brain tissue destruction at the same time might explain some of the

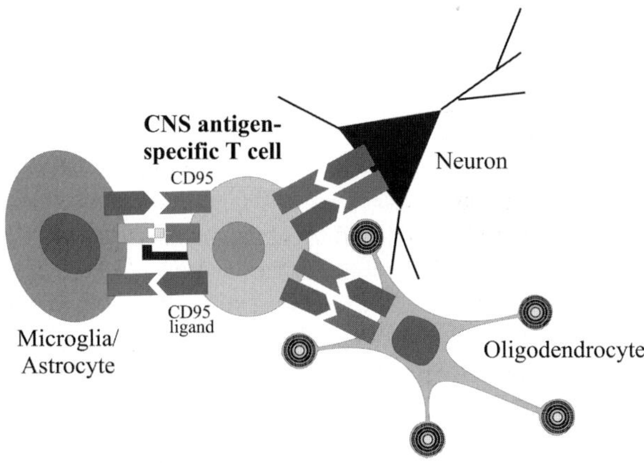

Fig. 1. The role for the CD95 system in neuroinflammation. The scenario might be that activated CNS-specific T cells transmigrate and recognize their specific antigen on antigen-presenting cells such as, e.g., microglia or astrocytes (Weber et al. 1994; Williams et al. 1994). Additionally, in an inflamed context neurons have been shown to be capable of presenting antigen to MHC-I-restricted $CD8^+$ T cells in vitro (Medana et al. 2000). Upregulating the death-mediating CD95 ligand on their surface upon T-cell-receptor mediated activation, T cells might subsequently interact with CD95-expressing targets such as oligodendrocytes and neurons leading to the tissue damage observed in neuroinflammatory diseases. On the other hand, the same antigen-specific T cells have been found to express the CD95 receptor, which might contribute to their apoptotic clearance at the end of the inflammatory CNS attack by ligation with the CD95 ligand. The latter was shown to be expressed on astrocytes and neurons in normal brain (Bechmann et al. 1999) and on microglial cells (D'Souza et al. 1996) as well as oligodendrocytes (Dowling et al. 1996) in inflamed CNS tissue

unexpected data on death receptor/death ligand systems. In particular, the failure of TNF-targeted strategies blocking the TNF activity in clinical MS trials might be due to the dual role of TNF-induced cell death in brain damage and defense (van Oosten et al. 1996; The Lenercept Multiple Sclerosis Study Group and The University of British Columbia MS/MRI Analysis Group 1999; Liu et al. 1999). Concerning the ubiquitously expressed CD95/CD95 ligand system, it has already been elucidated from the MS animal model that a blockade of the receptor or the ligand is a far-too-simple therapeutic intervention. Although autoimmune CNS disease is not a feature of mouse strains mutant for the CD95 or CD95 ligand genes, the course of EAE is different from that in wildtype mice. According to reports in active EAE induced by immunization with an encephalitogenic peptide, both CD95 and CD95-ligand-deficient mice present with attenuated disease (Malipiero et al. 1997; Sabelko et al. 1997; Waldner et al. 1997). However, CD95-deficient mice of a different genetic background displayed an unmodified course of the acute phase of active EAE but failed to remit (Suvannavejh et al. 2000). Contradictory results have also been reported regarding the amount of apoptotic T cells in the CNS and the extent of tissue damage in these CD95/CD95-ligand-deficient mouse models. In a different EAE model induced by transfer of encephalitogenic T cells (passive EAE), CD95-ligand-deficient mice developed enhanced clinical signs when immunized with wildtype T cells (Sabelko-Downes et al. 1999), indicating the role of CD95-mediated apoptosis in the regulation of T cells in EAE. Finally, a recent study using DNA vaccination against CD95 ligand prior to induction of active EAE in rats supported the dual role of the CD95 ligand in the course of EAE (Wildbaum et al. 2000). Thus, the protection from EAE by inhibition of the ligand in the initiation phase is most likely caused by a blockade of the T-cell-mediated tissue destruction via the CD95 system. The lack of remission resulting from the same intervention was explained by the fact that at this stage the damage processes can only be stopped by clearance of expanded T cells from the brain which involves CD95-mediated T-cell death (Fig. 1).

Imbalance of pro-apoptotic and anti-apoptotic factors might be a key mechanism in autoimmunity, in particular neuroinflammation. Downstream of death ligand/receptor interactions, the main candidates for apoptosis regulation are members of the bcl-2-family. The bcl-2 protein itself was first discovered in B-cell lymphoma, where translocation results in overexpression and overproduction (Hockenbery et al. 1990). This overproduction leads to increased cell survival. These studies identified bcl-2 as the "survival factor." Following that, a series of bcl-2 homologous proteins was discovered that possess both anti- and pro-apoptotic functions. To these belong the bax-protein and bcl-x, which arises through alternative splicing of the cell death-inducing protein bcl-x_S or the -inhibiting protein bcl-x_L (Oltvai et al. 1993; Boise et al. 1993). The ratio of death-agonists and -antagonists determines whether or not a cell dies (Reed 1994; Yang and Korsmeyer 1996). In MS plaques, high levels of TNF-α are expressed when active demyelination is observed (Bitsch et al. 2000), whereas expression of the anti-apoptotic Bcl-2 in oligodendrocytes correlates with remyelination (Kuhlmann et al. 1999). In the animal model of MS, overexpression of the apoptosis-inhibiting molecule bcl-x_L in T cells resulted in an earlier onset and a more chronic form of the disease presumably due to a reduced clearance of T cells from the brain (Issazadeh et al. 2000). Moreover, the anti-apoptotic molecule FLIP was reported to be overexpressed in T cells of MS patients (Semra et al. 2001). Our data on increased production of apoptosis blocking soluble CD95 receptor in the serum of patients with relapsing remitting MS further indicate that a dysbalance of apoptosis regulating factors contributes to the pathogenesis of the disease (Zipp et al. 1998a,b). Compatible with these findings, Zang et al. found increased frequencies of myelin-specific T cells in MS patients but not in healthy controls upon blockade of the CD95/CD95 ligand system which prevents in vivo activated T cells from undergoing apoptosis when again stimulated ex vivo (Zang et al. 1999). These data indicate an impairment of activation-induced death of myelin-specific T cells in MS patients.

12.3 The Role of the TRAIL System in Neuroinflammation

TRAIL is a newly identified member of the TNF family (Wiley et al. 1995), with CD95L being the most closely related. TRAIL can interact with five different receptors. Of these, only TRAIL receptor 1 (also referred to as DR4) (Pan et al. 1997b) and TRAIL receptor 2 (DR 5) (Sheridan et al. 1997) are capable of transmitting a death signal, whereas the transmembrane TRAIL receptor 3 (DcR1) and TRAIL receptor 4 (DcR2) do not contain a functional death domain (Degli-Esposti et al. 1997a,b). The latter two are considered to act as decoy receptors since their overexpression was shown to inhibit TRAIL-induced apoptosis (Pan et al. 1997a; Sheridan et al. 1997). The soluble receptor osteoprotegerin is reported to block TRAIL-mediated apoptosis by competitive inhibition of TRAIL binding to TRAIL receptor 1 and 2 (Emery et al. 1998).

Despite homologies, there are great differences between the TRAIL system on the one hand, and the TNF and CD95 system on the other hand. Previous studies emphasized a certain selectivity of TRAIL to induce apoptotic cell death in tumor cells (Ashkenazi et al. 1999; Walczak et al. 1999). However, this view has been challenged since soluble TRAIL was also reported to induce massive apoptosis in normal human hepatocytes (Jo et al. 2000) and human brain cells (Nitsch et al. 2000). The latter observation suggests a major role for this system in glial/neuronal damage.

Interestingly, human T cells despite bearing the death inducing TRAIL receptors 1 and 2 are resistant to TRAIL-mediated apoptosis (Wendling et al. 2000). Thus, TRAIL does not appear to be involved in the maintenance of peripheral tolerance by means of clonal T-cell deletion. TRAIL was, however, reported to be upregulated upon T-cell activation and in peripheral immune cells of patients with MS (Huang et al. 2000; Wendling et al. 2000). Recent studies in animal models of MS and rheumatoid arthritis suggested an inhibitory influence of TRAIL on T-cell growth and effector function (Song et al. 2000; Hilliard et al. 2001), whereas reversed signaling by TRAIL was shown to enhance T-cell proliferation (Chou et al. 2001). In the human immune system, soluble TRAIL was reported to inhibit T-cell proliferation and IFN-γ production in the absence of apoptotic cell death (Lünemann et al. 2001). These data indicate that TRAIL might act as a negative regulator of human T-cell function via mechanisms different from the induction apoptosis.

Thus, the TRAIL system does not appear to participate in the dual role of apoptotic mechanisms in neuroinflammation and might therefore serve as a particularly suitable target for therapeutic interventions because of its proposed differential role in T-cell control (T cells are resistant to TRAIL-mediated apoptosis) and CNS tissue damage (CNS cells are susceptible to TRAIL-mediated apoptosis). Ongoing research may lead to the development of a protective blockade of T-cell-mediated glial/neuronal cell death without abrogating the death-receptor-mediated removal of potentially harmful T cells (Aktas et al. 2001).

12.4 The Role of Apoptosis in Established Therapies in MS

Already established T-cell targeting therapies in MS, e.g., glucocorticoids and IFN-β, involve T-cell apoptosis in different ways. Glucocorticoids directly induce apoptosis in antigen-specific T cells, but subsequently protect these cells from CD95-mediated apoptosis (Zipp et al. 2000b). This observation might explain the lack of relevant long-term effect of glucocorticoids in the treatment of MS. The soluble CD95 receptor is not influenced by high-dose intravenous methylprednisolone pulse therapy (Zipp et al. 2000b).

There are contradictory results regarding the apoptosis properties of IFN-β in vitro. This immunomodulatory agent was shown to induce apoptosis (Nagatani et al. 1998; Kaser et al. 1999) and to rescue cells from apoptosis (Pilling et al. 1999). Several recent observations indicate that IFN-β modulates the CD95/CD95 ligand system. During IFN-β therapy, neutralizing IFN-β antibodies can be generated and may be related to treatment failure. In longitudinally studied patients, an increase of soluble CD95 correlated with early neutralizing IFN-β antibody production and clinical worsening (Zipp et al. 1998b). IFN-β treatment of MS patients results in a biphasic response, with an initial increase and subsequent decrease of serum CD95 levels (Zipp et al. 1998b). This kind of initial elevation was still seen after 6 months by other authors (Rep et al. 1999). In parallel, the latter study also reported an increase in CD95-expressing T cells during the first months of IFN-β therapy. We can confirm a moderate upregulation of CD95 cell surface expression after exposure to IFN-β (Zipp et al. 2000a), which has been reported in some cell types (Rep et al. 1999). Further, IFN-β increased mRNA and cell-surface expression of CD95 in leukemic cells (Selleri et al. 1997) and enhanced CD95 ligand-induced apoptosis of human malignant glioma cells (Roth et al. 1998). In conclusion, it is conceivable that IFN-β promotes T-cell apoptosis via unknown pathways, thus, reducing exacerbations in the disease course. Accordingly, we found an increase in the apoptosis rate of peripheral immune cells upon IFN-β therapy (Gniadeck et al. 2000). However, we were able to show that IFN-β does not induce apoptosis in T cells in a direct manner (Zipp et al. 2000a).

Acknowledgements. Original work was supported by the Gemeinnützige Hertie-Stiftung, the Deutsche Forschungsgemeinschaft to F. Zipp (ZI 448/7-1).

References

Aktas O, Osmanova V, Beyer M, Brocke S, Zipp F (2001) Therapeutic modulation of the TRAIL system in autoimmune CNS inflammation. J Neuroimmunol 118/1:52

Ashkenazi A, RC Pai, S Fong, S Leung, DA Lawrence, SA Marsters, C Blackie, L Chang, AE McMurtrey, A Hebert, L DeForge, IL Koumenis, D Lewis, L Harris, J Bussiere, H Koeppen, Z Shahrokh, RH Schwall (1999) Safety and antitumor activity of recombinant soluble Apo2 ligand. J Clin Invest 104:155

Bachmann R, Eugster HP, Frei K, Fontana A, Lassmann H (1999) Impairment of TNF-receptor-1 signaling but not fas signaling diminishes T-cell apoptosis in myelin oligodendrocyte glycoprotein peptide-induced chronic demyelinating autoimmune encephalomyelitis in mice. Am J Pathol 154:1417–1422

Bechmann I, Mor G, Nilsen J, Eliza M, Nitsch R, Naftolin F (1999) FasL (CD95L, Apo1L) is expressed in the normal rat and human brain: evidence for the existence of an immunological brain barrier. Glia 27:62–74

Bechmann I, Lossau S, Steiner B, Mor G, Gimsa U, Nitsch R (2000) Reactive astrocytes upregulate fas (CD95) and fas ligand (CD95L) expression but do not undergo programmed cell death during the course of anterograde degeneration. Glia 32:25–41

Bitsch A, Kuhlmann T, da Costa C, Bunkowski S, Polak T, Brück W (2000) Tumour necrosis factor alpha mRNA expression in early multiple sclerosis lesions: correlation with demyelinating activity and oligodendrocyte pathology. Glia 29:366–375

Boise LH, Gonzales-Garcia M, Postema CEDL, Lindsten T, Turka LA, Mao X, Nunez G, Thomson CB (1993) bcl-x, a bcl-2-related gene that functions as a dominant regulator of apoptotic cell death. Cell 74:597–608

Braun JS, Tuomanen EI (1999) Molecular mechanisms of brain damage in bacterial meningitis. Adv Ped Infect Dis 14:49–72

Braun JS, Novak R, Herzog K-H, Bodner SM, Cleveland JC, Tuomanen EI (1999) Neuroprotection by a caspase inhibitor in pneumococcal meningitis. Nat Med 5:298–302

Brunner T, Mogil RJ, LaFace D, Yoo NJ, Mahboubi A, Echeverri F, Martin SJ, Force WR, Lynch DH, Ware CF, et al (1995) Cell-autonomous Fas (CD95)/Fas-ligand interaction mediates activation-induced apoptosis in T-cell hybridomas. Nature 373:441–444

Chou AH, Tsai HF, Lin LL, Hsieh SL, Hsu PI, Hsu PN (2001) Enhanced proliferation and increased interferon-γ production by signal transduced through TNF-related apoptosis-inducing ligand. J Immunol 167:1347–1352

D'Souza SD, Bonetti B, Balasingam V, Cashman NR, Barker PA, Troutt AB, Raine CS, Antel JP (1996) Multiple sclerosis: Fas signaling in oligodendrocyte cell death. J Exp Med 184:2361–2370

Degli-Esposti MA, Dougall WC, Smolak PJ, Waugh JY, Smith CA, Goodwin RG (1997a) The novel receptor TRAIL-R4 induces NF-kappaB and protects against TRAIL-mediated apoptosis, yet retains an incomplete death domain. Immunity 7:813–820

Degli-Esposti MA, Smolak PJ, Walczak H, Waugh J, Huang C-P, DuBose RF, Goodwin RG, Smith CA (1997b) Cloning and characterization of TRAIL-R3, a novel member of the emerging TRAIL receptor family. J Exp Med 186:1165–1170

Dhein J, Walczak H, Baumler C, Decatin KM, Krammer PH (1995) Autocrine T-cell suicide mediated by APO-1(Fas/CD95). Nature 373:438–441

Dirnagl U, Iadecola C, Moskowitz MA (1999) Pathobiology of ischaemic stroke: an integrated view. Trends Neurosci 22:391–397

Dittel BN, Merchant RM, Janeway CA (1999) Evidence for Fas-dependent and Fas-independent mechanisms in the pathogenesis of experimental autoimmune encephalomyelitis. J Immunol 162:6392–6400

Dowling P, Shang G, Raval S, Menonna J, Cook S, Husar W (1996) Involvement of the CD95 (APO-1/Fas) Receptor/Ligand System in Multiple Sclerosis Brain. J Exp Med 184:1513–1518

Emery JG, McDonnell P, Burke MB, Deen KC, Lyn S, Silverman C, Dul E, Appelbaum ER, Eichman C, DiPrinzio R, Dodds RA, James IE, Rosenberg M, Lee JC, Young PR (1998) Osteoprotegerin is a receptor for the cytotoxic ligand TRAIL. J Biol Chem 273:14363

Ferguson B, MK Matyszak, MM Esiri, VH Perry (1997) Axonal damage in acute multiple sclerosis lesions. Brain 120:393–390

Flügel A, Schwaiger FW, Neumann H, Medana I, Willem M, Wekerle H, Kreutzberg GW, Graeber MB (2000) Neuronal FasL induces cell death of encephalitogenic T lymphocytes. Brain Pathol 10:353–364

Freyer D, Manz R, Ziegenhorn A, Weih M, Angstwurm K, Döcke WD, Meisel A, Schumann RR, Schönfelder G, Dirnagl U, Weber JR (1999) Cerebral endothelial cells release TNF-α after stimulation with cell walls of Streptococcus pneumoniae and regulate iNOS and ICAM-1 expression via autocrine loops. J Immunol 163:4308–4314

Gniadeck P, Aktas O, Claussnitzer A, Wendling U, Obert H, Zipp F (2000) Modulation of apoptosis in MS under therapy with interferon (IFN)-β1A. Rev Neurol 3S:76

Gold R, Hartung HP, Lassmann H (1997) T-cell apoptosis in autoimmune diseases: termination of inflammation in the nervous system and other sites with specialized immune-defense mechanisms. Trends Neurosci 20:399–404

Griffith TS, Herndon JM, Lima J, Kahn M, Ferguson TA (1995) The immune response and the eye. TCR alpha-chain related molecules regulate the systemic immunity to antigen presented in the eye. Int Immunol 7:1617–1625

Hilliard B, Wilmen A, Seidel C, Liu TS, Goke R, Chen Y (2001) Roles of TNF-Related Apoptosis-Inducing Ligand in Experimental Autoimmune Encephalomyelitis. J Immunol 166:1314–1319

Hisahara S, Araki T, Sugiyama F, Yagami KI, Suzuki M, Abe K, Yamamura KI, Miyazaki JI, Momoi T, Saruta T, Bernard CC, Okano H, Miura M (2000) Targeted expression of baculovirus p35 caspase inhibitor in oligodendrocytes protects mice against autoimmune-mediated demyelination. EMBO J 19:341–348

Hockenbery D, Nunez G, Milliman C, Schreiber RD, Korsmeyer SJ (1990) Bcl-2 is an inner mitochondrial membrane protein that blocks programmed cell death. Nature 348:334–336

Hohlfeld R, E Meinl, F Weber, F Zipp, S Schmidt, S Sotgiu, N Goebels, R Voltz, S Spuler, A Iglesias, H Wekerle (1995) The role of autoimmune T lymphocytes in the pathogenesis of multiple sclerosis. Neurology 45 (S6):33–38

Huang W-X, Huang MP, Gomes MA, Hillert J (2000) Apoptosis mediators fasL and TRAIL are upregulated in peripheral blood mononuclear cells in MS. Neurology 55:928–934

Issazadeh S, Abdallah K, Chitnis T, Chandraker A, Wells AD, Turka LA, Sayegh MH, Khoury SJ (2000) Role of passive T-cell death in chronic experimental autoimmune encephalomyelitis. J Clin Invest 105:1109–1116

Jo M, Kim TH, Seol DW, Esplen J, Dorko K, Billiar TR, Strom SC (2000) TNF-related apoptosis inducing ligand (TRAIL)-induced apoptosis in normal human hepatocytes. Nat Med 6:564–567

Karpus WJ, Ransohoff RM (1998) Chemokine regulation of experimental autoimmune encephalomyelitis: temporal and spatial expression patterns govern disease pathogenesis. J Immunol 161:2667–2671

Kaser A, Deisenhammer F, Berger T, Tilg H (1999) Interferon-beta 1b augments activation-induced T-cell death in multiple sclerosis patients. Lancet 353:1413–1414

Kornek B, Storch MK, Weissert R, Wallstroem E, Stefferl A, Olsson T, Linington C, Schmidbauer M, Lassmann H (2000) Multiple sclerosis and chronic autoimmune encephalomyelitis: a comparative quantitative study of axonal injury in active, inactive, and remyelinated lesions. Am J Pathol 157:267–276

Krammer PH (1998) CD95(APO-1/Fas)-mediated apoptosis: Live and let die. Adv Immunol 71:163–210

Krammer PH, Galle PR, Möller P, Debatin KM (1998) CD95 (APO-1/Fas)-mediated apoptosis in normal and malignant liver, colon, and hematopoetic cells. Ernst Schering Research Foundation, Academic Press

Kuhlmann T, Lucchinetti C, Zettl UK, Bitsch A, Lassmann H, Brück W (2000) Bcl-2-expressing oligodendrocytes in multiple sclerosis lesions. Glia 28:34–39

Leib SL, Kim YS, Chow LL, Sheldon RA, Tauber MG (1996) Reactive oxygen intermediates contribute to necrotic and apoptotic neuronal injury in an infant rat model of bacterial meningitis due to group B streptococci. J Clin Invest 98:2632–2639

Lenardo M, Chan KM, Hornung F, McFarland H, Siegel R, Wang J, Zheng L (1999) Mature T lymphocyte apoptosis-immune regulation in a dynamic and unpredictable antigenic environment. Annu Rev Immunol 17:221–253

Lenercept Multiple Sclerosis Study Group, The University of British Columbia MS/MRI Analysis Group (1999) TNF neutralization in MS: results of a randomized, placebo-controlled multicenter study. Neurology 53:457–465

Liu J, Marino MW, Wong G, Grail D, Dunn A, Bettadapura J, Slavin AJ, Old L, Bernard CC (1999) TNF is a potent anti-inflammatory cytokine in autoimmune-mediated demyelination. Nat Med 4:78–83

Lünemann JD, Waiczies S, Wendling U, Ehrlich S, Seeger B, Kamradt T, Zipp F (2001) Death ligand TRAIL inhibits proliferation of human (auto)antigen-specific T cells without inducing apoptosis or clonal anergy. J Neuroimmunol 118/1:101

Malipiero U, Frei K, Spanaus K-S, Agresti C, Lassmann H, Hahne M, Tschopp J, Eugster H-P, Fontana A (1997) Myelin oligodendrocyte glycoprotein-induced encephalomyelitis is chronic/relapsing in perforin knockout mice, but monophasic in Fas- and Fas ligand-deficient lpr and gld mice. Eur J Immunol 27:3151–3160

Martin R, McFarland HF, McFarlin DE (1992) Immunological aspects of demyelinating diseases. Annu Rev Immunol 10:153–187

Martino G, Hartung HP (1999) Immunopathogenesis of multiple sclerosis: the role of T cells. Curr Opin Neurol 12:309–321

Medana IM, Gallimore A, Oxenius A, Martinic MM, Wekerle H, Neumann H (2000) MHC class I-restricted killing of neurons by virus specific $CD8^+$ T lymphocytes is effected through the Fas/FasL, but not the perforin pathway. Eur J Immunol 30:3623–3633

Meyer R, Weissert R, Diem R, Storch MK, de Graaf KL, Kramer B, Bähr M (2001) Acute neuronal apoptosis in a rat model of multiple sclerosis. J Neuroscience 21:6214–6220

Nagatani T, Okazawa H, Kambara T, Satoh K, Nishiyama T, Tokura H, Yamada R, Nakajima H (1998) Effect of natural interferon-beta on the growth of melanoma cell lines. Melanoma Res 8:295–299

Nitsch R, Bechmann I, Deisz RA, Haas DLTN, Wendling U, Zipp F (2000) Massive cell death induced by tumor-necrosis factor-related apoptosis-inducing ligand (TRAIL) in adult human brain tissue. Lancet 356:827–828

Noseworthy JH, Lucchinetti C, Rodriguez M, Weinshenker BG (2000) Multiple Sclerosis. New Engl J Med 343:938–950

Oltvai ZN, Milliman ZL, Korsmeyer SJ (1993) Bcl-2 heterodimerizes in vivo with a conserved homolog, Bax, that accelerates programmed cell death. Cell 74:609–619

Pan G, Ni J, Wei Y, Yu G, Gentz R, Dixit VM (1997a) An antagonist decoy receptor and a death domain-containing receptor for TRAIL. Science 277:815–818

Pan G, O'Rourke K, Chinnaiyan AM, Gentz R, Ebner R, Ni J, Dixit VM (1997b) The Receptor for the Cytotoxic Ligand TRAIL. Science 276:111–113

Pender MP, Nguyen KB, McCombe PA, Kerr JF (1991) Apoptosis in the nervous system in experimental allergic encephalomyelitis. J Neurol Sci 104:81–87

Pilling D, Akbar AN, Girdlestone J, Orteu CH, Borthwick NJ, Amft, Scheel TD, Buckley CD, Salmon M (1999) Interferon-beta mediates stromal cell rescue of T cells from apoptosis. Eur J Immunol 29:1041–1050

Pitti RM, Marsters SA, Ruppert S, Donahue CJ, Moore A, Ashkenazi A (1996) Induction of apoptosis by Apo-2 ligand, a new member of the tumor necrosis factor cytokine family. J Biol Chem 271:12687–12690

Pouly S, Becher B, Blain M, Antel JP (2000) Interferon-γ modulates human oligodendrocyte susceptibility to Fas-mediated apoptosis. J Neuropathol Exp Neurol 59:280–286

Raine CS, Canella B (1992) Adhesion molecules and central nervous system inflammation. Semin Neurosci 4:201–211

Reed JC (1994) Bcl-2 and the regulation of programmed cell death. J Cell Biol 124:1–6

Rep MH, Schrijver HM, van Lopik T, Hintzen RQ, Roos MT, Ader HJ, Polman CH, van Lier RA (1999) Interferon (IFN)-beta treatment enhances CD95 and interleukin 10 expression but reduces interferon-gamma producing T cells in MS patients. J Neuroimmunol 96:92–100

Roth W, Wagenknecht B, Dichgans J, Weller M (1998) Interferon-alpha enhances CD95L-induced apoptosis of human malignant glioma cells. J Neuroimmunol 87:121–129

Roth W, Isenmann S, Naumann U, Kügler S, Bähr M, Dichgans J, Ashkenazi A, Weller M (1999) Locoregional Apo2L/TRAIL eradicates intracranial human malignant glioma xenografts in athymic mice in the absence of neurotoxicity. Biochem Bioph Res Co 265:479

Sabelko KA, Kelly KA, Nahm MH, Cross AH, Russell JH (1997) Fas and Fas ligand enhance the pathogenesis of experimental allergic encephalomyelitis, but are not essential for immune privilege in the central nervous system. J Immunol 159:3096–3099

Sabelko-Downes KA, Cross AH, Russell JH (1999) Dual role for Fas ligand in the initiation of and recovery from experimental allergic encephalomyelitis. J Exp Med 189:1195–1205

Schmied M, Breitschopf H, Gold R, Zischler H, Rothe G, Wekerle H, Lassmann H (1993) Apoptosis of T lymphocytes in experimental autoimmune encephalomyelitis. Evidence for programmed cell death as a mechanism to control inflammation in the brain. Am J Pathol 143:446–452

Schuchat A, Robinson K, Wenger JD, Harrison LH, Farley M, Reingold AL, Lefkowitz L, Perkins BA (1997) Bacterial meningitis in the United States in 1995. Active Surveillance Team. N Engl J Med 337:970–976

Schumann RR, Pfeil D, Freyer D, Buerger W, Lamping N, Kirschning CJ, Goebel UB, Weber JR (1998) Lipopolysaccharide and pneumococcal cell wall components activate the mitogen activated protein kinases (MAPK) erk-1, erk-2 and p38 in astrocytes. Glia 22:295–305

Selleri C, Sato T, Del Vecchio L, Luciano L, Barrett AJ, Rotoli B, Young NS, Maciejewski JP (1997) Involvement of Fas-mediated apoptosis in the inhibitory effects of interferon-a in chronic myelogenous leukemia. Blood 89:957–964

Semra YK, Seidi OA, Sharief MK (2001) Overexpression of the apoptosis inhibitor FLIP in T cells correlates with disease activity in multiple sclerosis. J Neuroimmunol 113:268–274

Sheridan JP, Marsters S, Pitti RM, Gurney A, Skubatch M, Baldwin D, Ramakrishnan L, Gray CL, Baker K, Wood WI, Goddard AD, Godowski P, Ashkenazi A (1997) Control of TRAIL-induced apoptosis by a family of signaling and decoy receptors. Science 277:818–821

Shevach EM (2000) Regulatory T cells in autoimmunity. Annu Rev Immunol 18:423–449

Smith T, Groom A, Zhu B, Turski L (2000) Autoimmune encephalomyelitis ameliorated by AMPA antagonists. Nat Med 6:62–66

Song K, Chen Y, Goke R, Wilmen A, Seidel C, Goke A, Hilliard B (2000) Tumor necrosis factor-related apoptosis-inducing ligand (TRAIL) is an inhibitor of autoimmune inflammation and cell cycle progression. J Exp Med 191:1095–1103

Suvannavejh GC, Dal Canto MC, Matis L, Miller SD (2000) Fas-mediated apoptosis in clinical remissions of relapsing experimental autoimmune encephalomyelitis. J Clin Invest 105:223–231

Sytwu HK, Liblau RS, McDevitt HO (1996) The roles of Fas/APO-1 (CD95) and TNF in antigen-induced programmed cell death in T cell receptor transgenic mice. Immunity 5:17–30

Tuomanen EI, Saukkonen K, Sande S, Cioffe C, Wright SD (1989) Reduction of inflammation, tissue damage, and mortality in bacterial meningitis in rabbits treated with monoclonal antibodies against adhesion-promoting receptors of leukocytes. J Exp Med 170:959–969

van Oosten BW, Barkhof F, Truyen L, Boringa JB, Bertelsmann FW, von Blomberg BM, Woody JN, Hartung HP, Polman CH (1996) Increased MRI activity and immune activation in two multiple sclerosis patients treated with the monoclonal anti-tumor necrosis factor antibody cA2. Neurology 47:1531–1534

van Parijs L, Peterson DA, Abbas AK (1998) The Fas/Fas ligand pathway and bcl-2 regulate T cell responses to model self and foreign antigens. Immunity 8:265–274

Walczak H, Degli Esposti MA, Johnson RS, Smolak PJ, Waugh JY, Boiani N, Timour MS, Gerhart MJ, Schooley KA, Smith CA, Goodwin R, Rauch CT (1997) TRAIL-R2: a novel apoptosis-mediating receptor for TRAIL. EMBO J 16:5386–5397

Walczak, H, Miller RE, Ariail K, Gliniak B, Griffith TS, Kubin M, Chin W, Jones J, Woodward A, Le T, Smith C, Smolak P, Goodwin RG, Rauch CT, Schuh JC, Lynch DH (1999) Tumoricidal activity of tumor necrosis factor-related apoptosis-inducing ligand in vivo. Nat Med 5:157

Waldner H, Sobel RA, Howard E, Kuchroo VK (1997) Fas- and FasL-deficient mice are resistant to induction of autoimmune encephalomyelitis. J Immunol 159:3100–3103

Weber F, Meinl E, Aloisi F, Nevinny-Stickel C, Albert E, Wekerle H, Hohlfeld R (1994) Human astrocytes are only partially competent antigen presenting cells. Possible implications for lesion development in multiple sclerosis. Brain 117:1323–1332

Wekerle H, Kojima K, Lannes-Vieira J, Lassmann H, Linington C (1994) Animal models. Ann Neurol 36:S47–53

Wendling U, Walczak H, Dörr J, Jaboci C, Weller M, Krammer PH, Zipp F (2000) Expression of TRAIL receptors in human autoreactive and foreign antigen-specific T cells. Cell Death Differ 7:637–644

Wildbaum G, Westermann J, Maor G, Karin N (2000) A targeted DNA vaccine encoding fas ligand defines its dual role in the regulation of experimental autoimmune encephalomyelitis. J Clin Invest 106:671–679

Wiley SR, Schooley K, Smolak PJ, Din WS, Huang CP, Nicholl JK, Sutherland GR, Smith TD, Rauch C, Smith CA, Goodwin RG (1995) Identification and characterization of a new member of the TNF family that induces apoptosis. Immunity 3:673–682

Williams K, Ulvestad E, Antel JP (1994) B7/BB-1 antigen expression on adult human microglia studied in vitro and in situ. Eur J Immunol 24:3031–3037

Woiciechowsky C, Asadullah K, Nestler D, Eberhardt B, Platzer C, Schoning B, Glockner F, Lanksch WR, Volk HD, Docke WD (1998) Sympathetic activation triggers systemic interleukin-10 release in immunodepression induced by brain injury. Nat Med 4:808–813

Yang E, Korsmeyer SJ (1996) Molecular thanatopsis: a discourse on the bcl-2 family and cell death. Blood 88:386–401

Zang YC, Kozovska MM, Hong J, Li S, Mann S, Killian JM, Rivera VM, Zhang JZ (1999) Impaired apoptotic deletion of myelin basic protein-reactive T cells in patients with multiple sclerosis. Eur J Immunol 29:1692–1700

Zipp F, Otzelberger K, Dichgans J, Martin R, Weller M (1998a) Serum CD95 of multiple sclerosis patients protects from CD95-mediated apoptosis. J Neuroimmunol 86:151–154

Zipp F, Weller M, Calabresi PA, Frank JA, Bash CN, Dichgans J, McFarland H, Martin R (1998b) Increased serum levels of soluble CD95 (APO1/Fas) in relapsing remitting multiple sclerosis. Ann Neurol 43:116–120

Zipp F, Krammer PH, Weller M (1999) Immune (dys)regulation in multiple sclerosis: role of the CD95/CD95 ligand system. Immunol Today 20:550–554

Zipp F, Beyer M, Gelderblom H, Wernet D, Zschenderlein R, Weller M (2000a) No induction of apoptosis by IFN-β in human. Neurology 54:524–526

Zipp F, Wendling U, Beyer M, Grieger U, Waiczies S, Wagenknecht B, Haas J, Weller M (2000b) Dual effect of glucocorticoids on apoptosis of human autoreactive and foreign antigen-specific T cells. J Neuroimmunol 110:214

Zysk G, Bruck W, Gerber J, Bruck Y, Prange HW, Nau R (1996) Anti-inflammatory treatment influences neuronal apoptotic cell death in the dentate gyrus in experimental pneumococcal meningitis. J Neuropathol Exp Neurol 55:722–728

Subject Index

(2′-5′) oligo A synthetase 93
2′5′-oligoadenylate synthetase 72

activation-induced cell death (AICD) 215
adhesion molecules 136
ageing 187
Aicardi-Goutières syndrome 67
Alzheimer's disease 15, 19, 97, 105, 113, 159, 161, 162, 166, 170, 183
amyloid 97
amyloidosis 97
anti-CD11b 137
anti-CD18 136
antigen-presenting cells 14, 18
apolipoprotein J 93
apoptosis 215
aquaporin-4 94
astrocytes 91, 138
ATP 109, 110
autoimmune encephalitis 187
axotomy 11, 15

β-amyloid 123
β-amyloid plaques 120
β-amyloid precursor protein 138
β_2-microglobulin 93
bcl-2-family 218
BLC 94
blood–brain barrier 6
BSE 85

calcification 66
cathepsin B 116
cathepsin S 92
CCL21 49, 51, 53, 54
CD11a-expression 139
CD11b/CD18 139
CD31 139
CD95 (APO-1/Fas) ligand 215
cellular adhesion molecules 68
cerebral ischemia 134
chemokines 68, 140
chemotaxis 30
chromogranin A 113
clinical trial 166, 170
clooxygenase-2 160
CNS 215
complement C1q 92
Cree encephalitis 67
cyclooxygenase-1 163, 167
cyclooxygenase-2 142, 163, 167
cytokine-induced neutrophil chemoattractant 144
cytokines 61, 91, 106, 140, 162

endothelial nitric oxide synthase (eNOS) 109
endothelin-1 140
enzyme-linked immunosorbent assay 86

excitotoxic injury 137
experimental autoimmune encephalomyelitis 64, 181, 215

facial nerve 183
Fas 111
FasL 111
focal cerebral ischemia 133
focal ischaemia 106
fractalkine 12, 17

glia activation markers 92
glial fibrillary acidic protein 62, 92
gliosis 91
global cerebral ischemia 136
glucose 111
glutamate 106, 110
glycoprotein-39 precursor 93

herpes simplex encephalitis 183
hippocampal sclerosis 182
HSP70 92
human trial 145
hypoglycaemia 111
hypoperfusion 136
hypoxia 111

ICAM-1 137, 138, 141
ICAM-2 138
IFN-α 66
IFN-β 221
IIGP protein 93
IL-1 16, 92, 140, 162
IL-10 143
IL-12 65
IL-1β 145
IL-3 64, 65
IL-4 143
IL-6 12, 16, 66, 92, 142, 162
immune privilege 216
inducible nitric oxide synthase 109, 142
inflammation 134
inflammatory 105
integrin 138
IP-10 94
ischaemic core 106
ischemia 14

Janus kinase 73

K^+ 109

L-selectin 138
leukocytes 134
LIM-homeodomain protein 7 (Lhx7) 94
lymphocytes 135
lymphocytic choriomeningitis virus 72
lysosomal-associated multi-transmembrane protein (LAPTm5) 94

Mac-1 139
macrophage inflammatory protein 44
macrophages 105, 135, 145
matrix metalloproteinases 68, 138
MCAO 53
meningitis 213
metalloproteases 123
metallothionein 93
methallothionein II 92
MHC class I 93
MHC class II 93
microglia 91, 105, 138, 145, 160, 161, 162, 164, 181
microglia cells 2
microglial activation 11, 12, 14, 16, 30, 105
microglial cultures 27, 38
migration 109

mitochondrial benzodiazepine receptor 180
mitochondrial permeability transition pore 116
monocyte chemoattractant protein 144
monocytes 135
multiple sclerosis 105, 181, 187, 193, 213
Mx-protein 93

neurodegeneration 97
neuroimmunology 45, 48
neuroinflammation 182
neuroinflammatory diseases 213
neuronal damage 54
neuroprotective 14
neutropenia 136
nitric oxide 106
no reflow phenomenon 135
nonsteroidal anti-inflammatory drugs (NSAIDs) 164, 166

Oligo A synthetase 95

P-selectin 138
p38 MAP kinase 117
p38 mitogen-activated protein kinase 168
p53 116
p55 141
p75 141
PAI-1 140, 141
Parkinson's disease 183, 187
PECAM-1 139
peri-infarct depolarizations 137
perineuronal 14
peripheral benzodiazepine binding site 180
perivascular monocyte 135
peroxisome proliferator-activated receptor-gamma 167, 170
phagocytosis 30, 108
PK11195 180
plasticity 186
positron emission tomography (PET) 179
potassium 110
prion 85
proteases 106

radioisotopes 179
reactive oxygen intermediates 112
receptors 12
reperfusion 134
RNA-dependent protein kinase 72

scrapie 85
sFasL 112
Shwartzman reaction 140
signal transducers and activators of transcription 73
SLC 53
statins 167
stroke 105, 133, 187
superoxide 106
suppressors of cytokine signaling 74

T cells 214
TGF-β-1 145
therapies 221
thromboxane A_2 140
tissue inhibitor of the matrix metalloproteinases 69, 70
tissue plasminogen activator (tPA) 137
tissue remodelling 186
TNF-related apoptosis inducing ligand (TRAIL/APO-2 ligand) 216
tolerance 215
transgenic 62
transgenic animals 165

transient ischemic attacks 136
transmissible spongiform
encephalopathies 85
tumor necrosis factor 12, 16, 65,
92, 110, 116, 140, 216

vasculitis 187
VCAM-1 138
vimentin 93
VLA-4 139

Wallerian-type degeneration 185

Ernst Schering Research Foundation Workshop

Editors: Günter Stock
Monika Lessl

Vol. 1 *(1991)*: Bioscience⇋Society – Workshop Report
Editors: D. J. Roy, B. E. Wynne, R. W. Old

Vol. 2 (1991): Round Table Discussion on Bioscience⇋Society
Editor: J. J. Cherfas

Vol. 3 (1991): Excitatory Amino Acids and Second Messenger Systems
Editors: V. I. Teichberg, L. Turski

Vol. 4 (1992): Spermatogenesis – Fertilization – Contraception
Editors: E. Nieschlag, U.-F. Habenicht

Vol. 5 (1992): Sex Steroids and the Cardiovascular System
Editors: P. Ramwell, G. Rubanyi, E. Schillinger

Vol. 6 (1993): Transgenic Animals as Model Systems for Human Diseases
Editors: E. F. Wagner, F. Theuring

Vol. 7 (1993): Basic Mechanisms Controlling Term and Preterm Birth
Editors: K. Chwalisz, R. E. Garfield

Vol. 8 (1994): Health Care 2010
Editors: C. Bezold, K. Knabner

Vol. 9 (1994): Sex Steroids and Bone
Editors: R. Ziegler, J. Pfeilschifter, M. Bräutigam

Vol. 10 (1994): Nongenotoxic Carcinogenesis
Editors: A. Cockburn, L. Smith

Vol. 11 (1994): Cell Culture in Pharmaceutical Research
Editors: N. E. Fusenig, H. Graf

Vol. 12 (1994): Interactions Between Adjuvants, Agrochemical and Target Organisms
Editors: P. J. Holloway, R. T. Rees, D. Stock

Vol. 13 (1994): Assessment of the Use of Single Cytochrome P450 Enzymes in Drug Research
Editors: M. R. Waterman, M. Hildebrand

Vol. 14 (1995): Apoptosis in Hormone-Dependent Cancers
Editors: M. Tenniswood, H. Michna

Vol. 15 (1995): Computer Aided Drug Design in Industrial Research
Editors: E. C. Herrmann, R. Franke

Vol. 16 (1995): Organ-Selective Actions of Steroid Hormones
Editors: D. T. Baird, G. Schütz, R. Krattenmacher

Vol. 17 (1996): Alzheimer's Disease
Editors: J.D. Turner, K. Beyreuther, F. Theuring

Vol. 18 (1997): The Endometrium as a Target for Contraception
Editors: H.M. Beier, M.J.K. Harper, K. Chwalisz

Vol. 19 (1997): EGF Receptor in Tumor Growth and Progression
Editors: R. B. Lichtner, R. N. Harkins

Vol. 20 (1997): Cellular Therapy
Editors: H. Wekerle, H. Graf, J.D. Turner

Vol. 21 (1997): Nitric Oxide, Cytochromes P 450, and Sexual Steroid Hormones
Editors: J.R. Lancaster, J.F. Parkinson

Vol. 22 (1997): Impact of Molecular Biology and New Technical Developments in Diagnostic Imaging
Editors: W. Semmler, M. Schwaiger

Vol. 23 (1998): Excitatory Amino Acids
Editors: P.H. Seeburg, I. Bresink, L. Turski

Vol. 24 (1998): Molecular Basis of Sex Hormone Receptor Function
Editors: H. Gronemeyer, U. Fuhrmann, K. Parczyk

Vol. 25 (1998): Novel Approaches to Treatment of Osteoporosis
Editors: R.G.G. Russell, T.M. Skerry, U. Kollenkirchen

Vol. 26 (1998): Recent Trends in Molecular Recognition
Editors: F. Diederich, H. Künzer

Vol. 27 (1998): Gene Therapy
Editors: R.E. Sobol, K.J. Scanlon, E. Nestaas, T. Strohmeyer

Vol. 28 (1999): Therapeutic Angiogenesis
Editors: J.A. Dormandy, W.P. Dole, G.M. Rubanyi

Vol. 29 (2000): Of Fish, Fly, Worm and Man
Editors: C. Nüsslein-Volhard, J. Krätzschmar

Vol. 30 (2000): Therapeutic Vaccination Therapy
Editors: P. Walden, W. Sterry, H. Hennekes

Vol. 31 (2000): Advances in Eicosanoid Research
Editors: C.N. Serhan, H.D. Perez

Vol. 32 (2000): The Role of Natural Products in Drug Discovery
Editors: J. Mulzer, R. Bohlmann

Vol. 33 (2001): Stem Cells from Cord Blood, In Utero Stem Cell Development, and Transplantation-Inclusive Gene Therapy
Editors: W. Holzgreve, M. Lessl

Vol. 34 (2001): Data Mining in Structural Biology
Editors: I. Schlichting, U. Egner

Vol. 35 (2002): Stem Cell Transplantation and Tissue Engineering
Editors: A. Haverich, H. Graf

Vol. 36 (2002): The Human Genome
Editors: A. Rosenthal, L. Vakalopoulou

Vol. 37 (2002): Pharmacokinetic Challenges in Drug Discovery
Editors: O. Pelkonen, A. Baumann, A. Reichel

Vol. 38 (2002): Bioinformatics and Genome Analysis
Editors: H.-W. Mewes, B. Weiss, H. Seidel

Vol. 39 (2002): Neuroinflammation – From Bench to Bedside
Editors: H. Kettenmann, G. A. Burton, U. Moenning

Supplement 1 (1994): Molecular and Cellular Endocrinology of the Testis
Editors: G. Verhoeven, U.-F. Habenicht

Supplement 2 (1997): Signal Transduction in Testicular Cells
Editors: V. Hansson, F. O. Levy, K. Taskén

Supplement 3 (1998): Testicular Function:
From Gene Expression to Genetic Manipulation
Editors: M. Stefanini, C. Boitani, M. Galdieri, R. Geremia, F. Palombi

Supplement 4 (2000): Hormone Replacement Therapy
and Osteoporosis
Editors: J. Kato, H. Minaguchi, Y. Nishino

Supplement 5 (1999): Interferon:
The Dawn of Recombinant Protein Drugs
Editors: J. Lindenmann, W.D. Schleuning

Supplement 6 (2000): Testis, Epididymis and Technologies
in the Year 2000
Editors: B. Jégou, C. Pineau, J. Saez

Supplement 7 (2001): New Concepts in Pathology and Treatment
of Autoimmune Disorders
Editors: P. Pozzilli, C. Pozzilli, J.-F. Kapp

Supplement 8 (2001): New Pharmacological Approaches
to Reproductive Health and Healthy Ageing
Editors: W.-K. Raff, M. F. Fathalla, F. Saad